Johannes Rabe

Lane-Precise Localization with Production Vehicle Sensors and Application to Augmented Reality Navigation

Schriftenreihe
Institut für Mess- und Regelungstechnik,
Karlsruher Institut für Technologie (KIT)

Band 042

Eine Übersicht aller bisher in dieser Schriftenreihe erschienenen Bände finden Sie am Ende des Buchs.

Lane-Precise Localization with Production Vehicle Sensors and Application to Augmented Reality Navigation

by

Johannes Rabe

KIT Scientific Publishing

Dissertation, Karlsruher Institut für Technologie
KIT-Fakultät für Maschinenbau

Tag der mündlichen Prüfung: 04. April 2018
Referenten: Prof. Dr.-Ing. Christoph Stiller
Prof. Dr. Philippe Bonnifait

Impressum

KIT Scientific Publishing

Karlsruher Institut für Technologie (KIT)
KIT Scientific Publishing
Straße am Forum 2
D-76131 Karlsruhe

KIT Scientific Publishing is a registered trademark
of Karlsruhe Institute of Technology.
Reprint using the book cover is not allowed.

www.ksp.kit.edu

Print on Demand 2018 – Gedruckt auf FSC-zertifiziertem Papier

ISSN 1613-4214
ISBN 978-3-7315-0854-0
DOI 10.5445/KSP/1000086154

Lane-Precise Localization with Production Vehicle Sensors and Application to Augmented Reality Navigation

Zur Erlangung des akademischen Grades

Doktor der Ingenieurwissenschaften

der Fakultät für Maschinenbau
des Karlsruher Instituts für Technologie (KIT)
genehmigte

Dissertation

von

JOHANNES RABE, M.SC.

aus Landsberg am Lech

Tag der mündlichen Prüfung: 04. April 2018
Hauptreferent: Prof. Dr.-Ing. Christoph Stiller
Korreferent: Prof. Dr. Philippe Bonnifait

Acknowledgement

This thesis is the result of my work in the team Augmented Reality at Daimler. I would like to thank the whole team for their support. Special thanks go to Dr. Marc Necker who gave me the opportunity for this work and Martin Hübner, Stephan Schmid, Dr. Janis Werner, Dr. Christian Grünler, and Sascha Quell for insightful discussions as well as collaboration for and proofreading of publications. Furthermore, I would like to thank the students I supervised for their work and cooperation.
I would like to thank Prof. Dr.-Ing. Christoph Stiller for his supervision and advise, but also for leaving me freedom during this work. Besides, I would like to thank Prof. Dr. Philippe Bonnifait for agreeing on bringing in his expertise as a correferee.
Special thanks go to my parents for supporting me in my decisions and, last but not least, to Cynthia for her patience and pressure to get this work done.

Unterschleißheim, June 2018 *Johannes Rabe*

Abstract

Next-generation automotive navigation systems will not only recommend certain lanes. Instead, they will also provide feedback on whether the driver is currently using a recommended lane or whether a lane-change is advisable. An especially intuitive visualization of this information can be achieved in an Augmented Reality (AR) navigation system. This requires a reliable ego-lane estimation. In this work, an algorithm for this purpose is developed, relying only on current production environment sensors. Furthermore, an analysis is performed on the required pieces for a lane-level AR route guidance system.

The proposed ego-lane estimation determines hypotheses for the current 2D-position and orientation of the vehicle by the means of a particle filter. By comparing the estimated position with a digital map, the current lane can be determined. Input data are Global Positioning System (GPS) position estimates, vehicle odometry, and data from environment sensors built into the used production vehicle such as front radar, stereo camera, and blind spot monitoring: The data is preprocessed in the respective electronic control units (ECUs) and contains information such as objects in front of the vehicle, detected lane-markings, and vehicles in the blind spot.

Two types of digital map are employed: A lanelet map describing the geometries of all lane boundaries and a current commercial navigation map that only contains road-level geometry. To apply measurements with tight likelihoods in situations with sparse particle sets, a combined weight update and sampling (CWUS) step is proposed that allows for reproducible results even when initializing the filter over large areas. Performance evaluations on more than 6 hours or 200 km of data from drives in urban, peri-urban, and highway scenarios show that the proposed algorithm reliably determines the current lane and that it is almost always available.

Besides localization, further topics on AR navigation are covered, such as the development of a possible system architecture. Furthermore, a soft map-matching applicable to AR applications is proposed and a heuristic that allows for lane-change recommendations even in situations with unclear map data that forbids lane-level routing or with uncertain localization.
Keywords: localization, augmented reality, navigation, route guidance, particle filter

Kurzfassung

Die nächste Generation von Navigationssystemen im Fahrzeug soll nicht nur bestimmte Fahrstreifen empfehlen, sondern dem Fahrer auch Rückmeldung geben, ob er sich aktuell auf dem richtigen Fahrstreifen befindet oder ob ein Fahrstreifenwechsel zu empfehlen ist. Besonders intuitiv lassen sich solche Empfehlungen in einem Augmented Reality (AR)-Navigationssystem darstellen. Zu diesem Zweck ist eine verlässliche Bestimmung des aktuell befahrenen Fahrstreifens notwendig. Inhalt dieser Arbeit ist der Entwurf eines Algorithmus zu diesem Zweck auf Basis aktueller Umgebungssensorik in einem Serienfahrzeug und die Analyse, welche Bestandteile für ein fahrstreifengenaues AR-Navigationssystem notwendig sind.

Der vorgestellte Algorithmus zur Fahrstreifenbestimmung verfolgt mehrere Hypothesen für die aktuelle 2D-Position und Orientierung des Fahrzeugs mithilfe eines Partikelfilters. Durch Vergleich der bestimmten Position mit einer digitalen Karte kann der aktuelle Fahrstreifen abgeleitet werden. Als Eingangsdaten werden neben Fahrzeugodometrie und Positionsschätzungen vom Global Positioning System (GPS) Daten von der im verwendeten Serienfahrzeug verbauten Umgebungssensorik verwendet: Die bereits in den entsprechenden Steuergeräten vorverarbeiteten Daten von Frontradar, Stereokamera und Totwinkelassistent beinhalten Informationen zu anderen Objekten vor dem Fahrzeug, erkannten Fahrstreifenmarkierungen und darüber, ob sich ein anderes Fahrzeug im toten Winkel befindet.

Als digitale Karte kommen sowohl eine Lanelet-Karte, die Geometrien aller Fahrstreifenränder beinhaltet, als auch eine aktuelle kommerzielle Navigationskarte, die lediglich Fahrbahngeometrien enthält, zum Einsatz. Um auch in Situationen mit geringer Partikeldichte noch Likelihoods von Messungen mit geringer Varianz anzuwenden, wird ein kombinierter Innovations- und Resamplingschritt vorgeschlagen. Dieser Schritt ermöglicht reproduzierbare Ergebnisse auch bei Initialisierung über einen größeren Bereich. In einer Auswertung auf mehr als 6 Stunden bzw. 200 km Daten von Fahrten in städtischen, kleinstädtischen und Autobahn-Szenarien kann gezeigt werden, dass der entworfene Algorithmus treffsicher den aktuellen Fahrstreifen bestimmt und fast immer verfügbar ist.

Neben der Lokalisierung werden Aspekte eines AR-Navigationssystems behandelt: Eine Systemarchitektur wird entworfen, außerdem wird ein weiches Map-Matching entwickelt, das für AR-Anwendungen geeignet ist. Mittels einer Heuristik können außerdem Fahrstreifenwechsel empfohlen werden, selbst wenn die verwendete Karte unvollständig ist und kein fahrstreifengenaues Routing möglich ist oder wenn die Lokalisierung sich unsicher ist.

Schlagwörter: Lokalisierung, Augmented Reality, Navigation, Routenführung, Partikelfilter

Contents

Notation and Symbols . . . . **xi**

1 Introduction **1**
1.1 Contribution 2
1.2 Structure 3

2 Fundamentals **5**
2.1 Digital Maps 5
2.1.1 Commercial Navigation Map 5
2.1.2 Lanelet Map 6
2.2 Sensor Data 7
2.2.1 Global Navigation Satellite System 8
2.2.2 Visual Lane-Marking Detection 8
2.2.3 Front Radar 9
2.2.4 Blind Spot Monitoring 9
2.2.5 Inertial Sensors 9
2.3 Dead Reckoning 10
2.4 Bayes Estimation 11
2.4.1 Overview 11
2.4.2 Kalman Filter 13
2.4.3 Extended Kalman Filter 14
2.4.4 Particle Filter 15

3 Localization **19**
3.1 Related Work 19
3.1.1 Absolute Positioning 20
3.1.2 Feature-based Localization 20
3.1.3 Localization on Vector Maps 21

3.2 Sensor Calibration 26
3.2.1 GPS Latency 27
3.2.2 Distance Estimate 28
3.2.3 Yaw Estimate 30
3.3 Preprocessing Radar Data 33
3.3.1 Moving Objects 34
3.3.2 Stationary Objects 36
3.4 Particle Filter Localization on Lanelet Map 36
3.4.1 System State 37
3.4.2 Initialization 39
3.4.3 Prediction 40
3.4.4 Map Update 41
3.4.5 Combined Weight Update and Sampling 43
3.4.6 Measurement Update 57
3.4.7 Resampling 60
3.4.8 Lane Hypothesis and Availability 61
3.5 Localization on Commercial Map 63
3.5.1 Preprocessing Map Data 63
3.5.2 Adaptations to Localization 69
3.5.3 Particle Filter Localization 74
3.5.4 Combination of Particle Filter and Extended Kalman Filter 76

4 Experimental Results 85
4.1 Evaluation Methods and Metrics 85
4.1.1 Test Tracks 85
4.1.2 Test Data 87
4.1.3 Ground Truth and Evaluation Logic 89
4.2 Particle Filter Localization on Lanelet Map 90
4.2.1 Simulation 90
4.2.2 Optimal System Design 100
4.2.3 Behavior in Different Scenarios 102
4.2.4 Convergence Behavior 105

4.2.5 Comparison of Proposed Method 107
4.2.6 Behavior Without Environment Sensors 112
4.3 Particle Filter Localization on Commercial Map 116
4.3.1 Evaluation Metrics 116
4.3.2 Quantitative Results 117
4.3.3 Critical Situations 118
4.4 Combination of Particle Filter and Extended Kalman Filter on Commercial Map 119
4.4.1 Quantitative Results 119
4.4.2 Critical Situations 120
4.5 Discussion 121

5 Application Example: Automotive AR Navigation 123
5.1 Related Work 123
5.2 System Architecture 126
5.3 Map-Matching for Automotive AR 128
5.3.1 Traditional Map-Matching 130
5.3.2 Allowed Driving Corridor 130
5.3.3 Offset Vector Estimation and Correction 131
5.3.4 Position Estimation 132
5.3.5 Results 132
5.4 Lane-Change Recommendations in Unclear Situations 134
5.4.1 Determining Possible Lane-Changes 135
5.4.2 Results 137

6 Conclusion and Outlook 139

References 143

Publications by the Author 161

Patent Applications 163

Supervised Theses 165

Notation and Symbols

List of Acronyms

ACC	Adaptive Cruise Control
ADAS	Advanced Driver Assistance System
AR	Augmented Reality
BSM	Blind Spot Monitoring
CWUS	combined weight update and sampling
DGPS	Differential GPS
DR	Dead Reckoning
ECU	electronic control unit
EKF	Extended Kalman Filter
EM	expectation maximization
ESC	Electronic Stability Control
GLONASS	Global Navigation Satellite System
GNSS	Global Navigation Satellite System
GPS	Global Positioning System
HMD	head-mounted display
HMI	human-machine interface
HUD	head-up display
IMM	interacting multiple model
IMU	inertial measurement unit
KF	Kalman filter
LDW	Lane Departure Warning
MAD	median absolute deviation
MMSE	minimum mean squared error
OGH	optimized geometric Hermite
OSM	OpenStreetMap

POI	point of interest
PF	particle filter
rpm	revolutions per minute
RTK	Real Time Kinematic
SLAM	simultaneous localization and mapping
VR	virtual reality

General Notation

Scalars	Regular (greek) lower case	a, b, c, σ, λ
Vectors	Bold (greek) lower case	$\boldsymbol{a}$, $\boldsymbol{b}$, $\boldsymbol{c}$, $\boldsymbol{\sigma}$, $\boldsymbol{\lambda}$
Matrices	Bold upper case	$\boldsymbol{A}$, $\boldsymbol{B}$, $\boldsymbol{C}$

System state estimation

$\boldsymbol{u}_k$	Control input at time t_k
$\boldsymbol{x}_k$	System state at time t_k
$\boldsymbol{x}_k^{[i]}$	System state hypothesis of particle i at time t_k
$\boldsymbol{z}_k$	Measurement observations at time t_k
v_k	Speed estimate at time t_k
$v_{\mathrm{c},k}$	Corrected speed estimate at time t_k
x_k	x-coordinate of state at time t_k
y_k	y-coordinate of state at time t_k
θ_k	Heading angle at time t_k
$\dot{\theta}_k$	Yaw rate at time t_k
μ_s	Mean of signal or random variable s
σ_s	Standard deviation of signal or random var. s
$bel\left(\boldsymbol{x}_k\right)$	Belief for state $\boldsymbol{x}_k$ after incorporating measurements $\boldsymbol{z}_k$
$\overline{bel}\left(\boldsymbol{x}_k\right)$	Belief for state $\boldsymbol{x}_k$ before incorporating measurements $\boldsymbol{z}_k$ ("intermediate belief", "prediction")

1 Introduction

In the 1990s, the US Global Positioning System (GPS) (XU 2007) and the Russian Global Navigation Satellite System (GLONASS) (HEIN 2000) became the first fully operational and widely used Global Navigation Satellite Systems (GNSSs). With their advent, positioning on the surface of Earth changed. While first GPS accuracy was artificially degraded by selective availability (HEIN 2000), the achievable positional accuracy increased by improved signal processing and the rise of techniques such as Real Time Kinematic (RTK) and Differential GPS (DGPS).

At the same time, in-vehicle navigation systems started to emerge that support the driver in finding his way to his destination. Initially, these systems relied on Dead Reckoning (DR) and only displayed the current position on a digital map such as the Honda Electro Gyrocator in 1981 (HONDA MOTOR CO. 2017). Actual route guidance was added two years later in Blaupunkt's EVA (B. SIMON 2017): After entering start and destination coordinates, the system calculated a possible route and guided by speech output. Due to the lack of storage space and the high effort of mapping, the covered area was quite limited.

With increasing storage space and map coverage, GNSS positioning, and data connections, automotive route guidance evolved to what it is now. The starting coordinate does not need to be entered by the user anymore, maps can cover groups of countries or continents, and the current delay due to traffic and possible detours can be displayed. As the level of detail of the used digital maps has grown, navigation systems with lane recommendation have made it to the market: When approaching complex urban intersections or highway interchanges, one or multiple lanes leading to the destination are highlighted.

These systems, however, do not determine whether the vehicle is currently driving on one of the recommended lanes or whether a lane-change might be advisable. Hence, lane-level localization is still task of the driver. While this

task seems trivial in many cases with two or three lanes, some drivers might be overstrained when first encountering complex urban intersections.
On other tracks, Augmented Reality (AR) systems have found their way from very technical applications such as wire bundle assembly (AZUMA 1997) to the mainstream. AR applications are ubiquitous (IKEA 2013; NIANTIC 2016) and also the automotive industry started to adopt (BMW GROUP 2017; MATHESON 2012; DAIMLER 2011). Besides entertainment and sales, also navigation systems can use AR to provide more intuitive ways of route guidance (MULLONI 2012; RAO et al. 2014b; SHEN 2013).
When implementing lane-level guidance in AR, it is not enough to merely know the recommended lanes. As all contents are displayed relative to the current position and orientation, the relative position to recommended lanes is required. This knowledge is equivalent to knowing the currently used lane and its geometry ahead of the vehicle. Hence, both lane-level AR route guidance and a traditional navigation system that provides lane-change recommendations depend on reliable knowledge of the ego-lane. With large urban intersections being one of the situations where a driver could benefit most from such a system, a localization algorithm should not fail there.

1.1 Contribution

In this work, such a localization algorithm is developed. The available input data is restricted to sensors readily available in a production vehicle. Based on a simple vector map rather than a feature- or landmark-based one, the algorithm is able to reliably determine the current lane in urban, suburban, and highway scenarios. A combined weight update and sampling step is proposed that combines benefits of analytical Bayes filters and particle approximations. This step allows for good localization accuracy and reproducible results while keeping the number of particles – and hence the required computational power – within reasonable bounds.
The developed algorithm is adapted to current commercial infotainment maps. The increased uncertainty in the map data is modeled by a set of heuristics in a map preprocessing step and the filter itself. Evaluation shows that lane-level

localization is possible on current maps in many situations. However, challenges remain when applying the solution to map data without lane-level geometry.
A system architecture for a navigation system that provides lane-change recommendations is developed. It allows for separate analysis of the three main system components localization, routing, and visualization. Aspects of an AR route guidance system related to localization, such as soft map-matching and lane-change recommendations on incomplete maps are further derived.

1.2 Structure

In the following Chapter 2, a short introduction to relevant fundamentals for this work, such as digital maps, the available sensor data, DR, and Bayes estimation theory and practice, is given. Then, the proposed localization method is developed in Chapter 3: after an overview of related work and the first sections of the filtering pipeline, a particle-filter based method for lane-precise localization on a lanelet map is presented, followed by two proposals for adapting it to current commercial maps. Chapter 4 shows the achievable localization performance of the proposed algorithms in both simulations and experiments and compares different design options, parameters, and the variants for different map data. The developed algorithms are then put into the context of an AR navigation system in Chapter 5 before Chapter 6 sums up the main findings and identifies some challenges for future development.

2 Fundamentals

In this chapter, relevant data and concepts are introduced. First, short introductions to the used types of digital map are given. The sensor data of the used test vehicle is described in Section 2.2 as it defines the constraints on the usable data within this work. Sections 2.3 and 2.4 give an overview on used DR and filtering methods that are employed in later sections.

2.1 Digital Maps

The digital maps used within this work are of two different types. One is a commercial navigation map as used for current automotive route guidance systems. The other one is a lanelet map, originally introduced in the context of research on automated driving (BENDER et al. 2014).

2.1.1 Commercial Navigation Map

Navigation maps try to describe the road network by a graph whose edges are denoted as *links* and connected at *nodes*. The geometry of roads is described via *shape points* which are an ordered list of coordinates corresponding to the center of the respective road. Each node connecting two links is a common shape point.

Additional information on the road can be given as link attributes. Among them are, for example, the number of lanes, speed limits, or functional road classes describing the role of the road in the road network. Typically, a link can only have one set of attributes. When, e.g., the speed limit changes, a link is replaced by two smaller links, each spanning the length of the respective attribute.

Lane attributes typically differ between basic maps, where only the number of through lanes without turn lanes is given and more detailed lane-level maps. In basic maps, some roads are even attributed with zero lanes. Lane-level maps also contain turn-, acceleration-, and deceleration lanes at situations that are considered difficult to navigate such as complex urban intersections and highway interchanges. At these situations, also lane connectivities can be provided that describe the connections between certain lanes before and after intersections. Furthermore, lane types, lane-marking types, and the information on whether a lane is forming or ending at a link can be attributed.

Current commercial navigation maps are made to allow routing and route guidance at a link-level together with a reasonable visualization of the road geometry and, partially, lane recommendations. This, however, leads to geometric inaccuracies in different situations: Intersections of directional roadbeds that are divided by physical dividers are modeled as tetragons and hardly describe the trajectory a vehicle would take over the intersection. At some intersections, turn lanes are divided by traffic islands from the rest of the intersection to allow quick turning and to give pedestrians a two-step way when passing the road. Such traffic islands are often only included in the map if at least one side has a certain length such as 25 m (HERE 2015) or 40 m (TomTom 2013). As the map geometry is used for visualization on a road-level, no explicit lane-level geometry is included.

2.1.2 Lanelet Map

Lanelet maps (Bender et al. 2014) have shown their applicability for highly automated driving (Althoff et al. 2017; Kuhnt et al. 2016; Ziegler et al. 2014) and provide a similar level of detail to other advanced digital maps such as the enhanced map (Betaille and Toledo-Moreo 2010), the Extended Digital Map (Nedevschi et al. 2013), or the local dynamic map (Mattern et al. 2010) used for vehicular localization (Selloum et al. 2009; Toledo-Moreo et al. 2010).

Furthermore, commercial map manufacturers work on comparable map formats containing lane geometries such as the HERE HD Live Map (Bonetti 2016) and the TomTom HAD Map and RoadDNA (TomTom 2016). Thus, it is reasonable to use a lanelet map for a lane-level localization solution.

The lanelet format is based on the OpenStreetMap (OSM) format. This format consists of three basic types of elements: *nodes*, *ways*, and *relations*. A node represents a coordinate, comparable to shape points in commercial maps. A way is an ordered lists of nodes, i.e. can describe polylines. A set of nodes and ways is described by a relation, where each entry can have a certain role. A relation can have a set of key-value-pairs that allow a user to interpret it as intended.
In a lanelet map, one lane or driving possibility is stored as a relation containing at least one way each as left and right boundaries. This relation is of type *lanelet* and the roles of the boundaries are *left* and *right*. Topology can be reconstructed by comparing left and right ways of a lanelet as well as first and last nodes of each boundary. The original lanelet map (BENDER et al. 2014) contains further elements such as regulatory elements that are not used within this work. In addition to the original lanelet format, the key *tunnel* with value *true* has been introduced to denote tunnel lanelets.
For this work, a lanelet map has been manually created based on satellite images. Its coverage is shown in Section 4.1.1.

2.2 Sensor Data

The sensor data used in this work is to a great extent defined by the built-in sensors and sensor data processing in the used test vehicle, a 2014 Mercedes-Benz S-Class (V222). The vehicle is equipped with the then-state-of-the-art driver assistance systems package Mercedes-Benz Intelligent Drive. This package contains, among others, Adaptive Cruise Control (ACC) ("DISTRONIC PLUS with Steering Assist and Stop&Go Pilot"), Lane Departure Warning (LDW) ("Active Lane Keeping Assist"), and Blind Spot Monitoring (BSM) ("Active Blind Spot Assist"). The functionality of these driver assistance systems is based on sensors such as a long range front radar, a stereo multi purpose camera, and short range radars, respectively. An overview of these and further sensors is depicted in Figure 2.1. Data from additional sensors such as the ultrasonic sensors used for the Active Parking Assist or the infrared camera enabling the Night View Assist are not used within this work. Internal sensors, such as wheel speed encoders can be accessed via the vehicle bus.

Additionally, a production level gyroscope, accelerometer, and GPS receiver are used. An overview of the sensor data available from these sensors is given in the following.

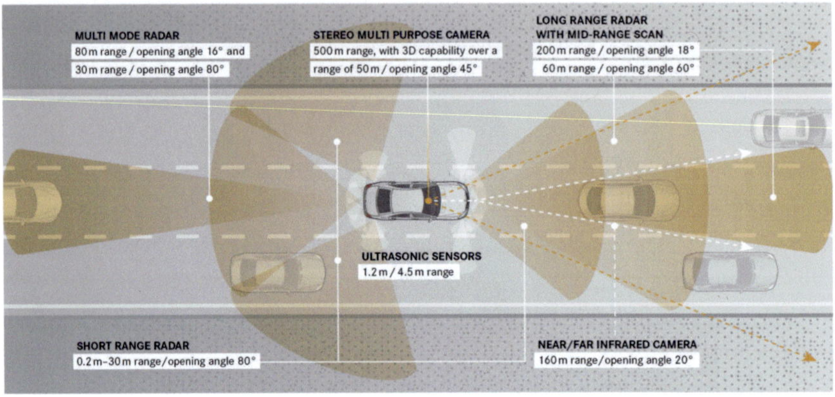

Figure 2.1: Sensors in 2014 Mercedes-Benz S-Class. Figure from DAIMLER 2012.

2.2.1 Global Navigation Satellite System

In the test vehicle, a u-blox 6 GPS receiver is used. The receiver uses speed measurements from the vehicle bus for an internal Dead Reckoning to provide accurate position estimates. The errors of the position estimate can be modeled as a bias that adds to a Gaussian noise component (CHAUSSE et al. 2005). In filters that allow non-Gaussian likelihoods, such as a particle filter, an explicit estimation of the bias vector can be avoided by approximating both errors as a cumulative uniform distribution.

2.2.2 Visual Lane-Marking Detection

Using video data from a stereo camera, lane-markings in front of the vehicle are detected. A clothoid model is used to describe the closest left and right boundary by their respective lateral distance, yaw angle between vehicle and marking, curvature, and change in curvature. Additionally, the type of marking (continuous, broken, block-marking etc.) is available. Discrete lane-change signals give hints on when a lane-marking is crossed by the vehicle – but need to be used with care as single- and double lane-changes cannot necessarily be distinguished.

2.2.3 Front Radar

The processed data from the front radar electronic control unit (ECU) contains discrete objects in front of the ego-vehicle. For each object, the relative distance, speed, and acceleration in vehicle x- and y-coordinates are available together with estimates for the respective standard deviations. Furthermore, an identification number, a value for the existence probability, a relative lane estimate (ego-lane, first/second to the left/right), and a classification into groups such as pedestrian, bicycle, car, truck, and guardrail are provided. No confidence information is available for the classification.

2.2.4 Blind Spot Monitoring

The short-range backward-facing radars provide information on the existence of vehicles in the blind spot of the vehicle. However, only the abstract information on whether the driver shall be warned is available instead of relative distance or velocity estimates.

2.2.5 Inertial Sensors

A set of sensors provide information on the movement of the vehicle. For all four wheels, wheel pulse counter values are available as well as the resulting revolutions per minute (rpm) and rotation direction. The vehicle speed estimated on these measurements is also provided.
The Electronic Stability Control (ESC) contains a yaw rate gyroscope and provides the unfiltered current yaw rate and an estimate on the offset of that value. The steering angle is available as the mean front wheel angle.
From a separate gyroscope, turn rates in three dimensions available. Its turn rates are calibrated in standstill using acceleration estimates from an accelerometer.

2.3 Dead Reckoning

DR describes the integration of inertial measurements such as speed, turn rate or wheel angle to determine a current position relative to a starting point (WALLENTOWITZ and REIF 2010). The simple point mass model used in the localization algorithm is shortly described here. It reduces the vehicle to a point mass for which a two-dimensional pose $\boldsymbol{x}_k = (x_k, y_k, \theta_k)$, consisting of a position x_k, y_k in a local coordinate frame and an orientation or heading θ_k relative to one of the coordinate axes is estimated. This point can move straight forward in heading direction, i.e. (ESKANDARIAN 2012, p. 408)

$$\begin{aligned} x_k &= x_{k-1} + \frac{2v_{k-1}}{\dot{\theta}_{k-1}} \sin\left(\frac{\dot{\theta}_{k-1}\Delta t_k}{2}\right) \cos\left(\theta_{k-1} + \frac{\dot{\theta}_{k-1}\Delta t_k}{2}\right) \\ &\approx x_{k-1} + v_{k-1}\Delta t_k \cos\left(\theta_{k-1}\right), \\ y_k &= y_{k-1} + \frac{2v_{k-1}}{\dot{\theta}_{k-1}} \sin\left(\frac{\dot{\theta}_{k-1}\Delta t_k}{2}\right) \sin\left(\theta_{k-1} + \frac{\dot{\theta}_{k-1}\Delta t_k}{2}\right) \\ &\approx y_{k-1} + v_{k-1}\Delta t_k \sin\left(\theta_{k-1}\right), \end{aligned} \tag{2.1}$$

given a speed v_k, yaw rate $\dot{\theta}_k$, and time difference $\Delta t_k = t_k - t_{k-1}$ at time instance k. Accordingly, the current heading θ_k can be determined based on the previous estimate θ_{k-1} and a yaw rate $\dot{\theta}_k$ as (ESKANDARIAN 2012, p. 408)

$$\theta_k = \theta_{k-1} + \Delta t_k \dot{\theta}_k, \tag{2.2}$$

describing a rotation of the vehicle around its vertical axis.

When using observed speeds $\hat{v}_k$ and yaw rate $\hat{\dot{\theta}}_k$, the respective ideal values can be replaced in Equations (2.1) and (2.2) by

$$\begin{aligned} v_k &= \hat{v}_k + \eta_k, \text{and} \\ \dot{\theta}_k &= \hat{\dot{\theta}}_k + \nu_k \end{aligned} \tag{2.3}$$

to include process noise values η_k and ν_k (ESKANDARIAN 2012, p. 409).

It must be noted that a more detailed vehicle model such as the bicycle model (RAJAMANI 2012) allows for a more correct modeling of the vehicle movement, especially in curves. However, the point mass model provides a reasonable approximation and requires little knowledge of the vehicle dimensions.

2.4 Bayes Estimation

In this section, a short overview of the principle of recursive Bayes estimation is given. Then, its special cases Kalman filter (KF), Extended Kalman Filter (EKF), and particle filter (PF) – which will be used later in this work – are introduced. For a detailed introduction to the topic, the reader may refer to THRUN et al. 2005 whose notation is partially adapted in this work. An exhaustive overview of Bayesian Filters is also provided in CHEN 2003.

2.4.1 Overview

Given a generic state-space model in discrete time, as shown in Figure 2.2, the goal is to estimate the current non-observable system state $\boldsymbol{x}_k \in \mathbb{R}^{N_x}, k \in \mathbb{N}$ based on the measurements (data) $\boldsymbol{z}_k \in \mathbb{R}^{N_z}, k \in \mathbb{N}$ for each time instance k. Besides measurements, also a control input $\boldsymbol{u}_k \in \mathbb{R}^{N_u}, k \in \mathbb{N}$ may be available. Within this work, no explicit control input is given. Instead, measurements that describe movement and rotation of the system state, are used in the role of control input as proposed by THRUN et al. 2005.

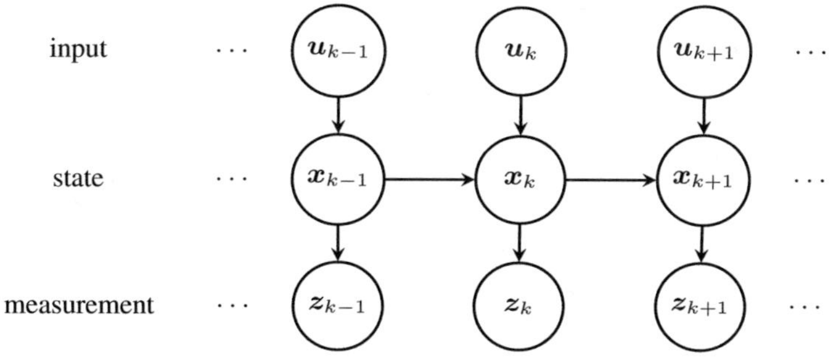

Figure 2.2: Bayes network, inspired by CHEN 2003.

The system can hence be modeled by two equations

$$\boldsymbol{x}_k = \boldsymbol{f}\left(\boldsymbol{x}_{k-1}, \boldsymbol{u}_k, \boldsymbol{n}_k\right) \tag{2.4}$$

$$\boldsymbol{z}_k = \boldsymbol{h}\left(\boldsymbol{x}_k, \boldsymbol{v}_k\right) \tag{2.5}$$

for a state transition function $\boldsymbol{f} : \mathbb{R}^{N_x} \mapsto \mathbb{R}^{N_x}$, a measurement function $\boldsymbol{h} : \mathbb{R}^{N_x} \mapsto \mathbb{R}^{N_z}$, process noise $\boldsymbol{n}_k$, and measurement noise $\boldsymbol{v}_k$ (CHEN 2003). These equations already take into account that the state $\boldsymbol{x}_k$ is complete, i.e. information on past states does not provide additional information for the present and future states. This completeness is described by the Markov property in a probabilistic sense. Furthermore, the connections in the Bayes network show that the current state $\boldsymbol{x}_k$ directly depends only on the current control input $\boldsymbol{u}_k$ and the previous state $\boldsymbol{x}_{k-1}$ and the current measurement $\boldsymbol{z}_k$ only on the current state $\boldsymbol{x}_k$. These properties can be used to simplify some of the following distributions.

The *initial density* $p(\boldsymbol{x}_0)$ describes the knowledge of the system state at the beginning of the estimation. The *transition probability* from time instance $k-1$ to k can be described as

$$p(\boldsymbol{x}_k|\boldsymbol{x}_{0:k-1}, \boldsymbol{z}_{1:k-1}, \boldsymbol{u}_{1:k}) = p(\boldsymbol{x}_k|\boldsymbol{x}_{k-1}, \boldsymbol{u}_k), \tag{2.6}$$

where $\boldsymbol{y}_{k_1:k_2} = \boldsymbol{y}_{k_1}, \boldsymbol{y}_{k_1+1}, \ldots, \boldsymbol{y}_{k_2}$ refers to a sequence of a certain random variable $\boldsymbol{y}$. The *likelihood* for a certain measurement $\boldsymbol{z}_k$ becomes

$$p(\boldsymbol{z}_k|\boldsymbol{x}_{0:k}, \boldsymbol{z}_{1:k-1}, \boldsymbol{u}_{1:k}) = p(\boldsymbol{z}_k|\boldsymbol{x}_k). \tag{2.7}$$

This allows for the estimation of the *posterior probability* or *belief*

$$bel(\boldsymbol{x}_k) = p(\boldsymbol{x}_k|\boldsymbol{z}_{1:k}, \boldsymbol{u}_{1:k}) \tag{2.8}$$

by repeated application of the Bayes formula

$$\begin{aligned} p(\boldsymbol{x}_k|\boldsymbol{z}_{1:k}, \boldsymbol{u}_{1:k}) &= \frac{p(\boldsymbol{z}_k|\boldsymbol{x}_k, \boldsymbol{z}_{1:k-1}, \boldsymbol{u}_{1:k})\, p(\boldsymbol{x}_k|\boldsymbol{z}_{1:k-1}, \boldsymbol{u}_{1:k})}{p(\boldsymbol{z}_k|\boldsymbol{z}_{1:k-1}, \boldsymbol{u}_{1:k})} \\ &= \eta\, p(\boldsymbol{z}_k|\boldsymbol{x}_k, \boldsymbol{z}_{1:k-1}, \boldsymbol{u}_{1:k})\, p(\boldsymbol{x}_k|\boldsymbol{z}_{1:k-1}, \boldsymbol{u}_{1:k}) \end{aligned} \tag{2.9}$$

for a normalization factor η equal for all $\boldsymbol{x}_k$ (THRUN et al. 2005). Employing the conditional independence

$$p(\boldsymbol{z}_k|\boldsymbol{x}_k, \boldsymbol{z}_{1:k-1}, \boldsymbol{u}_{1:k}) = p(\boldsymbol{z}_k|\boldsymbol{x}_k), \tag{2.10}$$

defining the *intermediate belief* $\overline{bel}(\boldsymbol{x}_k)$ before the incorporation of measurements from time instance k, and using the theorem of total probability and the Markov property

$$
\begin{aligned}
\overline{bel}\left(\boldsymbol{x}_k\right) &= p\left(\boldsymbol{x}_k | \boldsymbol{z}_{1:k-1}, \boldsymbol{u}_{1:k}\right) \\
&= \int p\left(\boldsymbol{x}_k | \boldsymbol{x}_{k-1}, \boldsymbol{z}_{1:k-1}, \boldsymbol{u}_{1:k}\right) p\left(\boldsymbol{x}_{k-1} | \boldsymbol{z}_{1:k-1}, \boldsymbol{u}_{1:k}\right) \mathrm{d}\boldsymbol{x}_{k-1} \\
&= \int p\left(\boldsymbol{x}_k | \boldsymbol{x}_{k-1}, \boldsymbol{u}_k\right) p\left(\boldsymbol{x}_{k-1} | \boldsymbol{z}_{1:k-1}, \boldsymbol{u}_{1:k-1}\right) \mathrm{d}\boldsymbol{x}_{k-1} \\
&= \int p\left(\boldsymbol{x}_k | \boldsymbol{x}_{k-1}, \boldsymbol{u}_k\right) bel\left(\boldsymbol{x}_{k-1}\right) \mathrm{d}\boldsymbol{x}_{k-1}
\end{aligned}
\tag{2.11}
$$

allows for writing this equation as (THRUN et al. 2005)

$$
bel\left(\boldsymbol{x}_k\right) = \eta p\left(\boldsymbol{z}_k | \boldsymbol{x}_k\right) \overline{bel}\left(\boldsymbol{x}_k\right). \tag{2.12}
$$

The general Bayes filter consists of recursively estimating the belief and intermediate belief, starting with an initial assumption $bel\left(\boldsymbol{x}_0\right) = p\left(\boldsymbol{x}_0\right)$.
The analytical evaluation of these equations is difficult or impossible for general probability distributions that may or may not have an analytic description. Hence, inference schemes have been developed that either solve the problem for special cases or approximate the solution on a grid-basis or by random sampling. In the following, first the Kalman filter is described which allows for an optimal solution for linear systems with Gaussian noise. Then, the extended Kalman filter expands the set of problems to nonlinear cases. Finally, the particle filter is introduced which allows for inference on arbitrary systems with arbitrary noise functions using Monte Carlo methods, i.e. random sampling.

2.4.2 Kalman Filter

To allow for the use of a Kalman filter (KALMAN 1960), the dynamic system must be modeled by linear equations, i.e. Equations (2.4) and (2.5) become

$$
\boldsymbol{x}_k = \boldsymbol{F}\boldsymbol{x}_{k-1} + \boldsymbol{G}\boldsymbol{u}_k + \boldsymbol{n}_k \tag{2.13}
$$

$$
\boldsymbol{z}_k = \boldsymbol{H}\boldsymbol{x}_k + \boldsymbol{v}_k \tag{2.14}
$$

with matrices $\boldsymbol{F}$, $\boldsymbol{G}$, $\boldsymbol{H}$ of respective dimensions and $\boldsymbol{n}_k$, $\boldsymbol{v}_k$ zero-mean Gaussian with corresponding covariance matrices $\boldsymbol{R}$ and $\boldsymbol{Q}$, respectively.
Within the filter, the state is represented as a Gaussian and described by its mean $\boldsymbol{\mu}_k$ and covariance $\boldsymbol{\Sigma}_k$. The estimation of the intermediate belief is denoted as

prediction and performed by plugging the mean of the previous state estimate $\boldsymbol{\mu}_{k-1}$ and the current control input $\boldsymbol{u}_k$ into the state transition function (THRUN et al. 2005)

$$\overline{\boldsymbol{\mu}}_k = \boldsymbol{F}\boldsymbol{\mu}_{k-1} + \boldsymbol{G}\boldsymbol{u}_k. \tag{2.15}$$

The covariance matrix is predicted as

$$\overline{\boldsymbol{\Sigma}}_k = \boldsymbol{F}\boldsymbol{\Sigma}_{k-1}\boldsymbol{F}^{\mathsf{T}} + \boldsymbol{R}. \tag{2.16}$$

In the *innovation* or *measurement update* step, the Kalman gain is determined based on the covariance prediction and the measurement noise as

$$\boldsymbol{K}_k = \overline{\boldsymbol{\Sigma}}_k\boldsymbol{H}^{\mathsf{T}}\left(\boldsymbol{H}\overline{\boldsymbol{\Sigma}}_k\boldsymbol{H}^{\mathsf{T}} + \boldsymbol{Q}\right)^{-1}. \tag{2.17}$$

This gain is then used to determine the new mean and covariance estimates (THRUN et al. 2005)

$$\boldsymbol{\mu}_k = \overline{\boldsymbol{\mu}}_k + \boldsymbol{K}_k\left(\boldsymbol{z}_k - \boldsymbol{H}\overline{\boldsymbol{\mu}}_k\right), \tag{2.18}$$

$$\boldsymbol{\Sigma}_k = \left(\boldsymbol{I} - \boldsymbol{K}_k\boldsymbol{H}\right)\overline{\boldsymbol{\Sigma}}_k. \tag{2.19}$$

2.4.3 Extended Kalman Filter

The KF can be extended to nonlinear systems with a dynamic system model as in Equations (2.4) and (2.5) by linearization up to the first term of the Taylor expansion. The Jacobians $\boldsymbol{J}_{F,k}$ and $\boldsymbol{J}_{H,k}$ are determined as

$$\boldsymbol{J}_{F,k} = \left.\frac{\partial \boldsymbol{f}\left(\boldsymbol{x}_{k-1}, \boldsymbol{u}_k, \boldsymbol{n}_k\right)}{\partial \boldsymbol{x}_{k-1}}\right|_{\boldsymbol{x}_{k-1}=\boldsymbol{\mu}_{k-1}} \tag{2.20}$$

$$\boldsymbol{J}_{H,k} = \left.\frac{\partial \boldsymbol{h}\left(\boldsymbol{x}_k, \boldsymbol{v}_k\right)}{\partial \boldsymbol{x}_k}\right|_{\boldsymbol{x}_k=\overline{\boldsymbol{\mu}}_k}. \tag{2.21}$$

The prediction and innovation of the estimated mean are performed using the nonlinear functions $\boldsymbol{f}\left(\boldsymbol{\mu}_{k-1}, \boldsymbol{u}_k, 0\right)$ and $\boldsymbol{h}\left(\overline{\boldsymbol{\mu}}_k, 0\right)$, respectively. In the calculation of intermediate covariance, Kalman gain, and posterior covariance, the Jacobians $\boldsymbol{J}_{F,k}$ and $\boldsymbol{J}_{H,k}$ are used like the system matrices $\boldsymbol{F}$ and $\boldsymbol{H}$, respectively. (THRUN et al. 2005)

2.4.4 Particle Filter

Unfortunately, many problems are neither linear nor is it possible to describe all relevant probabilities as unimodal Gaussians. In such cases, it is in general difficult or impossible to find an analytic description of the posterior distribution. It is, however, possible to approximate the distribution by a set of random samples, also called *particles*. This group of methods is referred to as particle filter or sequential Monte Carlo methods – inspired by physical random number generators such as dies and roulettes used in casinos (CHEN 2003). Other names for methods of this type include bootstrap filtering (GORDON et al. 1993) or condensation (ISARD and BLAKE 1998).

A PF iteratively follows the same sequence of steps as a KF or EKF. However, the steps are not performed for two parameters that describe the complete belief but rather on a multitude of samples. Whilst the covariance is not directly estimated in the filter, an *importance weight* is assigned to each particle and modified in the measurement update step.

Initially, a set of n samples is drawn from $p(\boldsymbol{x}_0)$, each particle i is assigned the same importance weight $w_0^{[i]} = \frac{1}{n}$ (ARULAMPALAM et al. 2002)[1]. This initial distribution can represent an initial guess for the system state or be uniformly distributed over the whole state space.

The prediction step is similar to the one in the EKF: Each particle i is transformed following the transition function (THRUN et al. 2005)

$$\boldsymbol{x}_k^{[i]} = \boldsymbol{f}\left(\boldsymbol{x}_{k-1}^{[i]}, \boldsymbol{u}_k, \boldsymbol{n}_k^{[i]}\right) \tag{2.22}$$

where $\boldsymbol{x}_{k-1}^{[i]}$ is the state of particle i and $\boldsymbol{n}_k^{[i]}$ a realization of the random variable $\boldsymbol{n}_k$. That means that a random component modeling the process noise of the state transition is added to every particle's state in the prediction step. This increases the covariance of the sample set similar to the increase in the covariance in a Kalman filter.

[1] The actual value of the weight is irrelevant as long as it is the same for all particles. The importance weights can later be normalized to sum up to 1.

In the update step, the likelihood function $p\left(\boldsymbol{z}_k \middle| \boldsymbol{x}_k^{[i]}\right)$ is evaluated for each sensor measurement $\boldsymbol{z}_k$ and particle i at its respective position in the state space and multiplied onto its importance weight (THRUN et al. 2005)

$$w_k^{[i]} = w_{k-1}^{[i]} p\left(\boldsymbol{z}_k \middle| \boldsymbol{x}_k^{[i]}\right) . \tag{2.23}$$

This equation is simplified for the case of one measurement per time instance k. Of course, it can be easily extended to situations with asynchronous measurements that lead to multiple or no measurements in a single step by multiplying the likelihood values for each measurement to an overall importance weight factor.

Over time, the weights of few particles tend to grow while most will converge toward zero. This effect is referred to as *particle* or *weight degeneracy* (DOUC and CAPPÉ 2005). One option to compensate this effect is to perform a *resampling*, i.e. to draw new samples from the existing particle set, depending on their accumulated importance weight. Multiple algorithms exist for this purpose and differ on how the new samples are drawn. An overview is given in DOUC and CAPPÉ 2005. For example, in multinomial resampling, the particles are kept in order and the cumulative weights are determined. Then, n random numbers are drawn from a uniform distribution over $[0, 1]$ and for each number r the particle i is chosen that fulfills

$$i = \arg\min_i \left(\sum_{j=0}^{i} w^{[j]} \geq r \right) . \tag{2.24}$$

With this approach, it can happen that the same particle is drawn n times even though it does not carry the complete weight of the set. In systematic or low-variance resampling, only one random number q is drawn from a uniform distribution over $\left[0, \frac{1}{n}\right]$ and n new particles are drawn that fulfill

$$i = \arg\min_i \left(\sum_{j=0}^{i} w^{[j]} \geq q + \frac{m}{n} \right) \forall m \in \{0, 1, \ldots, n-1\} . \tag{2.25}$$

In resampling, however, it must be taken care of that particle degeneracy is not replaced by *sample impoverishment*, the situation where there is no diversity among the particles (ARULAMPALAM et al. 2002).

Some types of particle filter can be proven to converge to the approximated posterior distribution for large particle numbers $n \to \infty$ under certain constraints (DOUCET et al. 2001). However, the possible number of particles is limited by the computational capabilities of the used machine and the complexity of likelihood evaluations. When lane-level localization methods based on a particle filter are presented in the next chapter, one point is to consider this and to keep at least the most relevant hypotheses – i.e. particles – alive.

3 Localization

With these fundamentals at hand, it is now possible to design methods for lane-precise localization. After an overview of related work, the first sections of the filtering pipeline are developed: Calibration of inertial sensor data (Section 3.2) and preprocessing radar data (Section 3.3). A particle filter approach for lane-precise localization is developed and described in detail in Section 3.4. This approach is then adapted to commercial maps in two variants that are presented in Section 3.5.

3.1 Related Work

Within this chapter, methods for lane-precise localization of a road vehicle using a given set of sensor data and different digital maps are developed. In this field, a notable number of approaches have been published. As this thesis lies in the context of AR applications, one might think of using algorithms known in the virtual reality (VR)/AR community. However, these approaches mostly do not provide a localization relative to a digital road map. Relevant publications for the overall AR navigation system are reviewed in Section 5.1 while the focus within this section lies in approaches on vehicle localization. Three main branches exist: There is absolute positioning, where the goal is to accurately determine the position relative to a coordinate frame spanning Earth. Localization relative to digital maps can be split into methods relying on sensor-specific features and point landmarks, and others using pure geometric map data. A coarse overview of the first two branches is shown in the following. Then, a more detailed survey on the latter is given.

3.1.1 Absolute Positioning

Absolute positioning is nowadays widely based on Global Navigation Satellite System such as GPS, GLONASS, or Galileo. While the price for survey grade receivers with techniques such as RTK and DGPS has decreased over the last decades, they remain too expensive for the use in automotive navigation systems. DGPS allows for horizontal errors below 3 m when close to a base station (DU and BARTH 2008) and when fused with vehicle odometry, average errors below 1 m can be achieved (CARON et al. 2007).
Often, Bayes filters such as KF and its derivatives or PF are used to improve raw GNSS fixes by relative movement (CARON et al. 2007; EL NAJJAR and BONNIFAIT 2005; KRAKIWSKY et al. 1988). These approaches can be divided into loosely coupled methods, where GNSS position estimates are used (EL NAJJAR and BONNIFAIT 2005), and tightly coupled methods on the raw pseudorange measurements (Y. LI et al. 2006). Besides GNSS and inertial measurements, also map data can be used to improve the position estimate: In FOUQUE et al. 2008, an EKF includes heading information from a digital map to support the orientation estimate. Due to the possible offset of map data and the one-dimensional description of road links, no position information from the map are used.
When trying to determine the currently used lane, all positioning methods suffer from this drawback: Often, the absolute accuracy of current commercial digital maps is around 5 m. Hence, a centimeter-accuracy position may still not match with the correct lane or even road in the map. For accurate lane assignment, relative localization methods are required.

3.1.2 Feature-based Localization

A widespread approach to localization in research on automated driving is based on features or landmarks that may or may not be sensor-specific. In the basic concept, a feature map is generated in a mapping or reference drive. In the mapping drive, often high-precision GNSS setups are used (e.g. SCHREIBER et al. 2013). Alternatively, state-of-the-art simultaneous localization and mapping (SLAM) techniques allow for the generation of a map by solving an optimization problem (e.g. LATEGAHN and STILLER 2012).

The sensors used for environment perception vary: Some groups use a monocular camera (CHARMETTE et al. 2009; NODA et al. 2011), others employ stereo camera rigs (LATEGAHN and STILLER 2012; PINK 2008; SCHREIBER et al. 2013) or laser scanners (LEVINSON et al. 2007; QIN et al. 2012). Sometimes, the map is reduced to features that can be used by different sensors in the localization step (TESSIER et al. 2007). A different approach is shown in CORNICK et al. 2016: They employ a ground penetrating radar to generate an underground map and use this map for localization. Their motivation is that the underground changes its appearance less frequently than overground structures such as trees or parking cars – or just the illumination situation. However, research also tackles the problem of finding features that are robust against e.g. changes in illumination (LATEGAHN et al. 2013).
With the used sensor, also the used features and their source for the mapping process varies: In PINK 2008 and NODA et al. 2011, aerial or satellite images are used to extract features, e.g. lane-markings on the roads. Other features present in road environments are curbs and intersections (QIN et al. 2012).
In the localization step, different methods are used. Among these are different optimization methods such as least squares in graph-optimization (LATEGAHN and STILLER 2012), a modified iterative closest points approach (PINK 2008) or a direct least median of squares (NODA et al. 2011). Alternative approaches include particle filters (LEVINSON et al. 2007; QIN et al. 2012) and particle swarm optimization (CORNICK et al. 2016). LEVINSON et al. later change from a particle filter to a histogram filter and estimate the probability distribution over a grid in the surroundings of a current coarse position estimate (LEVINSON and THRUN 2010). Like this, they avoid situations where no particle is close to the actual position.

3.1.3 Localization on Vector Maps

The transition from feature- and landmark-based maps to pure vector maps is fluent. While lane-markings can be considered and used as landmarks, they also often determine the vector-based geometry of the lanes. One possible cut can be made at the level of detail of mapped lane-markings: If the exact position of each dash of broken lane-markings is available or stop lines and

arrow markings are mapped, one might consider it a feature map as e.g. in SCHREIBER et al. 2013. Similarly, lane-markings or lane boundaries may be mapped as one-dimensional lines (BENDER et al. 2014) or as two-dimensional areas with two boundaries themselves (TAO et al. 2013). Within this section, an overview of localization methods on maps with road or lane geometry is given.
One approach to determining the currently driven lane is based on absolute positioning: By estimating the current position as accurately as possible and comparing it to an accurate digital road map, the current lane can be found. This approach is used by DU et al. who use a low-cost DGPS receiver with a local base station and a lane-level road map constructed from aerial images. Through curve-to-curve matching of some of the last points onto the acquired lane centerlines, the current lane can be estimated. In extensive tests, they can correctly determine the current lane in 97% of the time. (DU and BARTH 2008)
Similarly, BAUER et al. estimate the current position with a many-dimensional particle filter whose system state contains, among others, the vehicle position and orientation, receiver clock error and drift, troposphere and ionosphere errors, and whether a line-of-sight connection to the tracked satellites is available. They can determine the ground truth lane in 96.5% of the evaluated 10 minutes data set after initialization from the ground truth position. (BAUER et al. 2016)
However, current road maps have limited absolute accuracy and any road link can be mapped up to 5 m away from its actual position. Furthermore, it is not always clear how lanes arrange around the mapped centerline of a road. A perfect absolute position is hence only of limited use in determining the currently used lane on current commercial maps.
Most other approaches employ some kind of relative localization. Often, other vehicles, infrastructure, and road-markings are used. Somewhere between absolute positioning and relative localization lies the work of DAO et al.: A group of vehicles determine their respective positions by combining GPS and DR. They communicate their position estimate to other vehicles and each vehicle can estimate the probabilities for which vehicle being on which of the available lanes. In simulations, they can in general determine the correct lane but sometimes lane-changes are detected too late (DAO et al. 2006). In real-world experiments, up to 95% correct lane estimates can be achieved for moving cars, while at walking speeds the success rate is even higher (DAO et al.

2007). Unfortunately, cooperative solutions to localization are not possible in a standalone navigation system.

Like for feature-based localization, Bayes theory is used in many approaches toward lane-level localization. Bayesian networks are used in POPESCU et al. 2012 and NEDEVSCHI et al. 2013. With a camera, a set of road features like lane-marking types and direction arrows are detected. An extended digital map contains lane-level information like lane-width, the type of markings and arrows, and the position of the centerline in approaches to intersections. Through inference the probabilities for different lanes is determined and can then be used as input to a particle filter algorithm. With the static Bayesian network, 5.6% of incorrectly identified lanes remain, with the following particle filter, the error rate can be reduced to 3.5% in an 8 km test set.

Alternatively, a discrete Bayes filter (histogram filter) can be employed as proposed by MAYER et al. 2014 and SVENSSON and SÖRSTEDT 2016. SVENSSON and SÖRSTEDT split the in-lane-localization and lane assignment into two separate problems. The relative positions and velocities of other vehicles are found by fusing radar and camera data and used to validate the lane hypotheses. A lane-change detection provides hints on whether the vehicle is still traveling on the same lane as in the previous time instance. No explicit evaluation on the recall ratio is given and the time-series analysis is still limited to two-lane scenarios in SVENSSON and SÖRSTEDT 2016.

Continuous state variables are possible when an EKF is used as in TAO et al. 2013: The authors fuse GPS and lane-markings from a camera image with a digital map that contains the inner boundaries of both lane-markings for each lane. Through the estimated covariance, an integrity value can be provided. In experiments, the absolute error was never larger than three standard deviations. On average, the position error was 0.26 m but at most 1.56 m which is considered lane-precise by the authors. It remains unclear how the system is initialized and what happens if the initialization is not within the correct lane.

MATTHAEI et al. also employ an EKF. In their work, they perform pose estimation relative to an occupancy and reflectance grid that were previously mapped using lidar and camera data. By including stationary structures in the map, the approach does not require lane-markings for lateral localization.

With the proposed method, the lateral error remains below 0.75 m most of the evaluated distance on an inner city main road (MATTHAEI et al. 2014). However, in this work, a lidar sensor should be avoided. Furthermore, the used grid maps are far more detailed than the intended vector maps.

Multiple hypotheses for the current lane or pose can be tracked with a particle filter as has been done for example in CHAUSSE et al. 2005. Their filter estimates the 2D-pose of a vehicle equipped with steering angle and speed sensors as well as a GPS receiver and a camera. Lane-markings detected by the camera are used in the update step to validate particle positions relative to a highly accurate digital map. They can achieve low standard deviations of the particle cloud in experiments and show how lane-changes and roundabouts can be handled.

A particle filter fusing only GNSS and DR with a digital map is shown in TOLEDO-MOREO et al. 2009: A GNSS update is performed depending on the quality of the current fix. The map data is incorporated in such that only positions on the road are valid. They achieve mean errors of 0.57 m when using Frenet coordinates but the horizontal positioning errors do not allow for reliable lane-level positioning – even provided a map with high absolute accuracy. This approach is extended by SELLOUM et al. 2009 to allow for direct use of Frenet coordinates in the prediction step. The position accuracy stays within the same region.

While CHAUSSE et al. 2005 and others compare the position to lane-markings with where they are detected in a camera image, MATTERN et al. 2010 follows an inverse approach: The map data is transformed into image hypotheses for each particle and the coherence between predicted and actual image is used to update importance weights. Their map features curbs, lane-markings where the actual position of dashes can be reconstructed, and stop lines retrieved from aerial images. Lane-level accuracy is proven by reaching a lateral position error of 0.35 m.

Similarly, VAN HAMME et al. 2013 only relies on image data from a monocular camera. Using visual odometry, the texture of the ground is extracted close to the vehicle trajectory. Then, linear sections are classified into road surface and not-road surface with a Gaussian mixture classifier. The position within the classified road surface provides hints on the currently used lane. In an evaluation of 32 segments of 120 m each, 94% correct lane assignments can be achieved if

the whole segment is used. By only using image data, the approach is limited by the camera aperture angle to roads with at most three lanes.
Another interesting method is proposed by SCHÜLE: Radar sensor measurements are integrated into a local radar map. As the position of the vehicle relative to the radar map is known, this local map can be matched with a digital road map to achieve a global pose estimate of the ego-vehicle. With that approach, the mean error can be reduced compared to pure GPS-based navigation (SCHÜLE et al. 2011). A local lane map based on camera data is proposed as an alternative to the radar map. It allows for similar accuracy to the radar map (SCHÜLE et al. 2013).
ECKEL employs particle filters for both airborne tracking and lane-level localization. For the latter, she estimates the current link and lane together with the longitudinal offset along the one-dimensional links. An EKF estimates the current absolute position. This position is used in the measurement update step to weight the particles as well as lane-marking types and lane-changes from a visual lane-marking detection. The evaluation focuses on road bifurcations and does not give insight on the overall performance of the algorithm. (SZOTTKA 2013)
When detecting not only the markings of the ego-lane but also neighboring lanes, these data can be used for localization as well. LU et al. 2014 shows an approach based on these and observed lane-changes to figure out the used lane of an OSM. In an evaluation on KITTI data, the lane selection is successful in around 78%. Reasons for errors are, among others, differences in the number of lanes between ground truth position and filtered position, and road barriers that are recognized as lane-markings.
Instead of lane-markings, KUHNT et al. employ objects detected by environmental sensors such as cameras or lidars. These objects are assumed to be traveling on the roads digitized in a lanelet map. When initializing the filter at the ground truth position, the absolute position error stays below 1.5 m for most of the evaluated 10 minute drive and the accuracy can be considered lane-level. (KUHNT et al. 2016).
An approach similar to SELLOUM et al. 2009; TOLEDO-MOREO et al. 2009 is used in F. LI et al. 2017: By mainly tracking possible positions through DR and validating them with GPS fixes, a position on a lane-level map shall be found.

The focus of the work lies in not only providing the correct lane but also integrity measures. In around 51% of the time, the correct lane is identified but the filter keeps multiple hypotheses which almost always include the correct lane.
A method not based on a specific filtering technique is proposed in C. R. SIMON 2012: Based on production sensors such as a camera and radar, environment models are estimated and used to determine the current lane. A line model uses up to eight lane-markings detected by a multi-purpose camera. The marking type allows for estimating the direction of travel of each modeled lane and the distance to the road boundary. Moving and standing vehicles are fused in an object model for the same purpose. In extensive experiments throughout Germany, the correct number of lanes can be determined in 75% of the time, the correct lane index in 85% of the time if road boundaries are detected. Including map data that contains the correct number of lanes allows to increase the ratio of correct lane assignments to 90%. Yet, the result in urban scenarios is not yet perfect: Almost 20% of the lane assignments are false within cities.
MEINKE 2015 proposes a method based on optimization. Trajectory pieces are fitted against centerlines of lanes in a lanelet map. The residuals of the least-squares minimization and penalties based on environment sensor data determine a score that is used to determine the current lane. While the achievable error rate is low, restricting evaluation to situations where one hypothesis stands out over the others reduces the system availability to around 60% of the time. (RABE et al. 2016a)

3.2 Sensor Calibration

For the use in the localization solution, a suitable source to drive the Dead Reckoning has to be chosen. For this purpose, the performance of the point mass model described in Section 2.3 has been evaluated for two different yaw rate estimates: one available on the vehicle bus from the ESC and one from a directly attached gyroscope. Their error properties are important for the DR component of the localization solution. Hence, the characteristics of the errors are described and methods for sensor calibration have been developed. Due to the lack of a high-precision inertial measurement unit (IMU) in the test vehicle,

the errors have been compared to the position and course estimates of a u-blox 6 GPS receiver with activated DR.
As the GPS receiver exhibits a notable latency that would distort estimates for distance and yaw, its latency is estimated first. Then, the distance estimate based on wheel speeds is characterized and the behavior of different vehicle models is evaluated. Calibration schemes for the velocity and yaw rate estimates are proposed.

3.2.1 GPS Latency

The used GPS receiver provides estimates for the current position, course, and speed at a frequency of 1 Hz. The speed estimates can be compared to speed estimates of other vehicle sensors and used to determine the latency between both values.
One option to do that is the maximization of the cross-correlation coefficient between the GPS speed estimate u and the vehicle speed from the speed control system or wheel speeds v, both available at 50 Hz. For this purpose, the high-frequency speed is downsampled by a factor of 50 after being shifted in time by the number of time steps p corresponding to the latency τ (RABE et al. 2016b):

$$\hat{p} = \arg\max_{p} \rho_{uv}(p) = \frac{\sum_{k=0}^{n} u_k v_{50k-p}}{\sqrt{\sum_{k=0}^{n} u_k^2 v_{50k-p}^2}}. \tag{3.1}$$

This can be used to estimate the latency as $\hat{\tau} = \hat{p} \cdot 0.02\,\mathrm{s}$ to be around $\hat{\tau} \approx 0.35 \ldots 0.4\,\mathrm{s}$. As data from different drives in 2015 and 2016 have been used, it can further be determined to be not changing much over time.
Another possibility is to use the travelled distance between two GPS fixes and a minimum mean squared error (MMSE) approach: Using the simple point mass vehicle model described in Section 2.3, position estimates $\boldsymbol{x}(t)$ relative to a starting point are created. These position estimates are first shifted in time by a possible latency τ and positions $\boldsymbol{x}(t_{\mathrm{GPS}} + \tau)$ at time instances corresponding to GPS fixes t_{GPS} are linearly interpolated. The distances

$$d_v^{\tau}(t_{\mathrm{GPS}}) = \left\| \boldsymbol{x}(t_{\mathrm{GPS}} + \tau) - \boldsymbol{x}(t_{\mathrm{GPS}} + T_{\mathrm{GPS}} + \tau) \right\| \tag{3.2}$$

between two interpolated positions and the corresponding distances between two GPS fixes $d_{\mathrm{GPS}}\left(t_{\mathrm{GPS}}\right)$ are determined and the respective mean squared error

$$MSE_{\tau} = \frac{1}{N}\sum_{1}^{N}\left(d_{v}^{\tau}\left(t_{\mathrm{GPS}}\right) - d_{\mathrm{GPS}}\left(t_{\mathrm{GPS}}\right)\right)^{2} \tag{3.3}$$

is calculated. The latency estimate $\hat{\tau}$ is the value τ that minimizes the mean squared error:

$$\hat{\tau} = \arg\min_{\tau}\left(\frac{1}{N}\sum_{1}^{N}\left(d_{v}^{\tau}\left(t_{\mathrm{GPS}}\right) - d_{\mathrm{GPS}}\left(t_{\mathrm{GPS}}\right)\right)^{2}\right). \tag{3.4}$$

This value can be estimated to be $\hat{\tau} \approx 0.4\,\mathrm{s}$ which is used throughout this work.

3.2.2 Distance Estimate

When using wheel speed sensors to estimate the distance travelled, several parameters play a role. Due to the vertical deflection of the tire changing for different wheel speeds, the effective rolling circumference changes with vehicle speed (RAJAMANI 2012), even though within tight boundaries (SCHRAMM et al. 2014). Additionally, tire air pressure, temperature, tire tread wear, and of course changing the wheels or tires have influence on the actual rolling circumference. To evaluate and model this influence, one can either use detailed models of tires and environment or estimate the behavior using GPS measurements.
Using the dynamic wheel circumference available from the vehicle bus as a starting point, the deviation of speed and distance estimates from their respective real values is estimated using the distances between two GPS fixes and distance estimates based on wheel speeds measurements, as in Equation (3.2). The evaluation is based on a data set of around 8.5 hours or 365 km.
From the deviation in estimated distance for different speeds in Figure 3.1, one can see that the speed-based distance is lower than the actual distance for higher speeds. This is caused by the used wheel circumference being smaller than the actual wheel circumference. The effect is increased for higher speeds by using a constant wheel circumference.

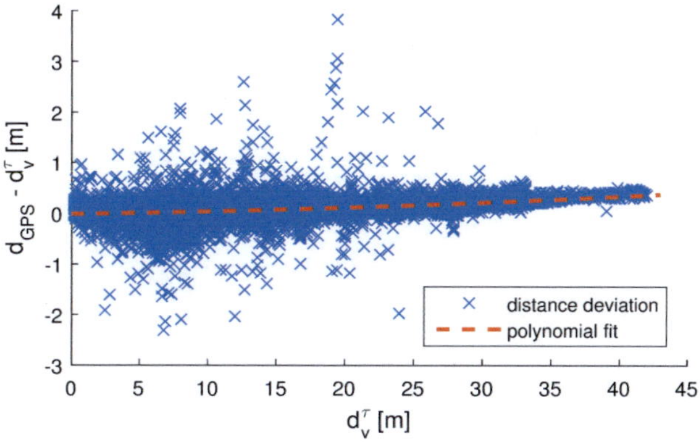

Figure 3.1: Deviation in estimated distance travelled in one second between GPS and wheel speeds using a constant wheel circumference of 2.15 m. Polynomial fit of the form $a(d_v^\mathcal{T})^2 + bd_v^\mathcal{T}$ depicted as dashed line.

Fitting a function of the form $a(d_v^\mathcal{T})^2 + bd_v^\mathcal{T}$ into the data in a least-squares sense yields the distance deviation

$$\Delta d = d_{\mathrm{GPS}} - d_v^\mathcal{T} \approx \frac{0.0001}{1\,\mathrm{m}}(d_v^\mathcal{T})^2 + 0.0041 d_v^\mathcal{T}. \tag{3.5}$$

To determine the behavior of the noise, this term is subtracted from the distance deviation, as shown in Figure 3.2. A slight increase in the deviation with higher speeds can be observed.

To determine the behavior over different speeds, the data is sorted into bins of 5 m/s width and the standard deviation is estimated for each bin, leading to the stairstep graph in Figure 3.2. It can be seen that the standard deviation varies between 0.04 m/s and 0.21 m/s but is neither decreasing or increasing monotonically with increasing speed. The increased standard deviations for the $[5\mathrm{m/s}, 10\mathrm{m/s})$ and $[15\mathrm{m/s}, 20\mathrm{m/s})$ bins is due to an increased number of outliers. These happen e.g. in curves because the GPS position estimate tends to cut corners in a different manner than the used vehicle model. The overall standard deviation for the used data set is around $\sigma_v = 0.15\,\mathrm{m/s}$.

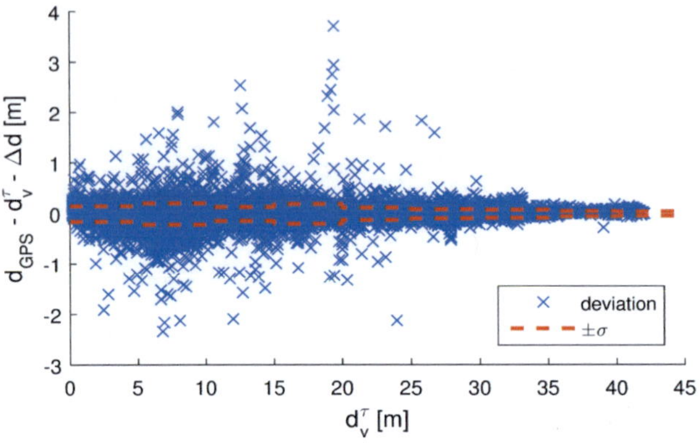

Figure 3.2: Deviation in estimated distance traveled in one second between GPS and wheel speeds corrected by the polynomial term in Equation (3.5). Standard deviation estimates for the 5 m/s bins shown as stairs.

3.2.3 Yaw Estimate

Two sources of yaw rate estimates are available. To choose one of them, it matters how much their values and properties differ. The vehicle provides an unfiltered and unadjusted yaw rate together with a current offset estimate via the vehicle bus. Additionally, a gyroscope is available that is calibrated by compensating its slowly varying component. The pose estimates based on these data sources are compared to the course and position estimates from the GPS receiver to estimate their drift over time.

To this end, the DR model was initialized at a GPS position estimate and then aggregated over time. The resulting position estimates for an exemplary log file depicted in Figure 3.3 give an impression of their error. After around 7 km through downtown and 19 km over highways, the position estimates based on both the gyroscope and ESC data differ noticeably from the GPS position estimates while the general shape of the trajectory can be recognized.

For the ESC yaw rate, the behavior with and without the offset estimate by the ECU is compared. Its characteristics can be seen in Figure 3.4: One can see that the estimates using the offset estimate (blue curve) exhibit noticeable changes over time.

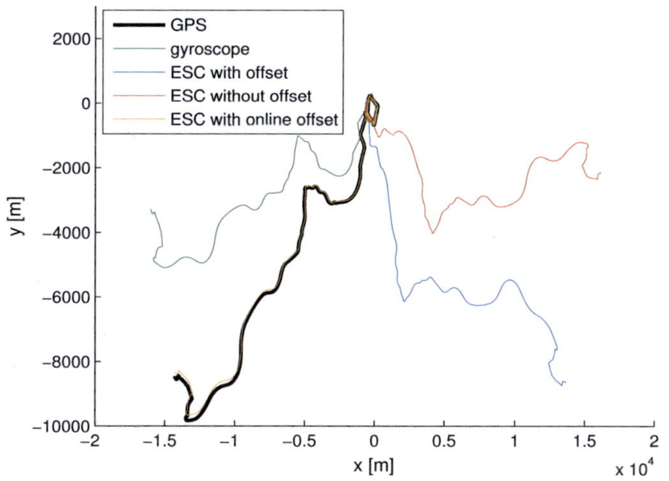

Figure 3.3: Evolution of position estimates of GPS and point mass vehicle model based on gyroscope and ESC yaw rate with different offset estimates.

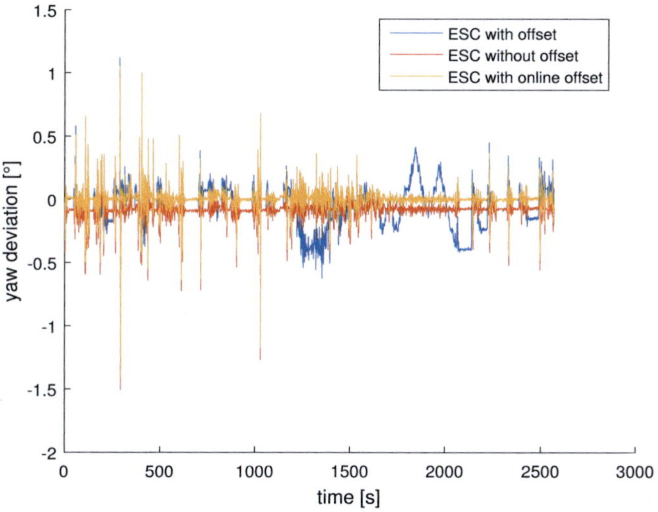

Figure 3.4: Deviation in estimated yaw difference in one second between GPS and point mass vehicle models using ESC yaw rate with and without ECU estimated offset, and with online estimated offset as described.

When ignoring this offset estimate (orange curve), the deviation remains rather constant – though biased – for the whole time. The observed bias was determined to stay around $-0.09\,^\circ/\mathrm{s}$ for most of the logged data. It seems promising to estimate the yaw rate offset online. For this purpose, it is interesting whether it changes over different yaw rate values.
No strong correlation between these two values is present, as can be seen in Figure 3.5. Hence, it is reasonable to only estimate a constant bias online. This is done by finding the median of the deviations of the course differences between the raw ESC yaw rate and the GPS course estimate. It is started as soon as 20 GPS measurements are available and determined over a window of at most 300 values.

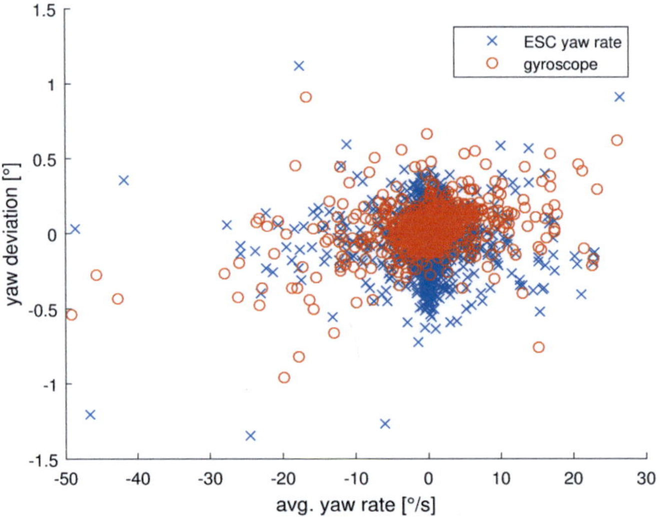

Figure 3.5: Deviation in estimate course difference in one second between GPS and point mass vehicle models based on ESC yaw rate and gyroscope over GPS change in course estimate.

With this approach, the yaw bias is compensated to a large extent, compare Figure 3.4, and the drift in the trajectory is kept within reasonable bounds, as seen in Figure 3.3: Even after around 26 km of driving, the estimated trajectory using the online ESC offset lies close to the GPS trace. To quantify the influence of this correction, outliers that may occur due to incorrect GPS course estimates when standing and similar situations are filtered out by first determining the median absolute deviation (MAD) and then removing all deviations that deviate from the median more than twice the estimated population standard deviation $\sigma = 1.4826 \cdot MAD$ (WILCOX 2001). In a second step, the sample standard deviation and mean are estimated from the reduced sample set.

The original yaw estimate based on the ESC yaw rate and the offset provided by the ECU has a mean of $\mu_{\dot{\theta},\text{offs}} = -0.009°/\text{s}$ and a standard deviation of $\sigma_{\dot{\theta},\text{offs}} = 0.111°/\text{s}$. Without this offset, a noticeable non-zero mean of $\mu_{\dot{\theta},\text{raw}} = -0.095°/\text{s}$, with a lower standard deviation of $\sigma_{\dot{\theta},\text{raw}} = 0.022°/\text{s}$ can be observed. When estimating the offset online and reducing it accordingly, the mean can be reduced to $\mu_{\dot{\theta},\text{onl}} = 0.0013°/\text{s}$ with a standard deviation of $\sigma_{\dot{\theta},\text{onl}} = 0.020°/\text{s}$.

The directly attached gyroscope is already calibrated by another system component using information from an accelerometer. The resulting data does not exhibit an almost constant offset as the ESC yaw rate does. Experiments with the proposed online calibration could not improve the statistics of the sensor readings. For this gyroscope, a mean deviation of $\mu_{\dot{\theta},\text{gyro}} = 0.021°/\text{s}$ with a standard deviation of $\sigma_{\dot{\theta},\text{gyro}} = 0.069°/\text{s}$ was determined. For these reasons, both the online offset-corrected ESC yaw rate and the gyroscope yaw rate are considered as data source for the yaw rate estimate.

3.3 Preprocessing Radar Data

Before fusing radar data into a localization algorithm, one should take a detailed look at it. Objects classified as vehicles might be infrastructure, reflexions, or in other ways not correctly identified environment. An example is shown in Figure 3.6, where a lot of single measurements and small object clusters can be seen along the bottom to top stretch in the figure. These objects are classified as vehicles but correspond to posts of a fence rather than actual vehicles that have driven to the left of the ego-vehicle. The only actual vehicles in this figure are the green and yellow segments at the top left end of the trajectory.

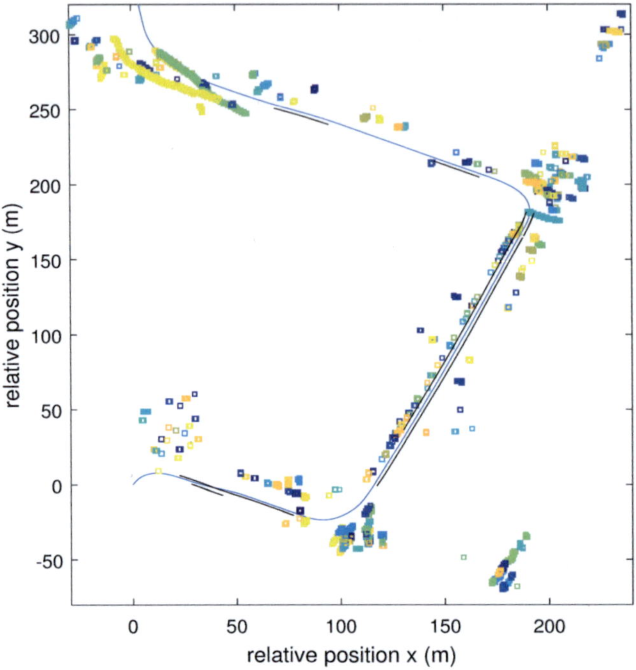

Figure 3.6: Radar objects with existence probability above 70% classified as cars and trucks over a trajectory of around 600 m (blue). Tracked objects with same id are depicted in the same color. Lane-markings detect by the front-facing camera are depicted in black to provide orientation.

A similar behavior can be seen upon examination of radar objects classified as guardrails. Furthermore, many stationary objects are classified as guardrails and hence one can find many vehicles standing on or besides the road e.g. when parking or waiting in a traffic jam among them. Therefore it is required to preprocess the radar data and filter out objects that do not represent interesting parts of the environment. This approach can also be described as a type of gating method known in target tracking (MITCHELL 2007).

3.3.1 Moving Objects

As many stationary objects are classified as cars and trucks by the radar, the main criterion to distinguish vehicles is their estimated velocity and actual movement.

Hence, objects are filtered in this manner: When an object with a certain id first appears, its position relative to the ego-vehicle is projected to a local coordinate frame and its initial position and velocity is stored. With the next measurement of this object, its new estimated position in local coordinates is determined and it is checked whether it is plausible given the estimated relative velocity and the traversed distance. If the new position is not plausible or too much time has passed since the last observation, the object is reset. Otherwise, the new position is stored along with the initial one and the distance between the newest and initial position is compared to two thresholds: If the total distance is above e.g. 8 m, the object is accepted as a vehicle. Otherwise, if the total distance is above e.g. 3 m but the object has already been observed many times, it can also be accepted. If the total distance is below both thresholds, the object is stored and the system waits for the next measurement of the same object.

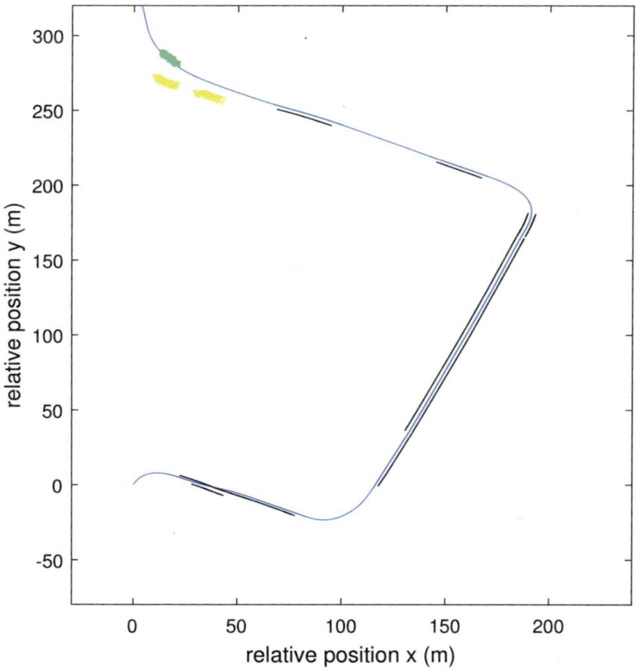

Figure 3.7: Radar objects classified as cars and trucks after filtering stage.

Additionally, objects further away from the ego-vehicle have proven less reliable and hence vehicles with a distance above 70 m from the ego-vehicle are not forwarded. This allows to track moving objects further away and use them as soon as they come closer. Objects that have been observed during high yaw rates of the ego-vehicle are discarded as well. The result can be seen in Figure 3.7 where no unwanted measurements remain and only the vehicles depicted in green and yellow at the top left situation are left.

3.3.2 Stationary Objects

In a similar manner, implausible stationary objects are sorted out. Only objects that occur multiple times around the same projected absolute position are accepted. As soon as an object is observed at a significantly different position, the observation count is reset. Again, objects further away than 70 m from the ego-vehicle are not forwarded but used for the aggregation step.

3.4 Particle Filter Localization on Lanelet Map

This section describes a particle filter-based approach to lane-precise localization on a lanelet map. After the description of the estimated system state, it leads through the algorithm as shown in Figure 3.8.
The particles are initialized as described in Section 3.4.2. To use two different input signals for the odometry prediction, two particle sets are used and their states are varied in two separate prediction steps (Section 3.4.3). Map information is incorporated in a map update (Section 3.4.4). The combined weight update and sampling step used for the distances to detected lane-markings is motivated and explained in Section 3.4.5 while the general measurement update step is presented in Section 3.4.6. Section 3.4.7 shows the used resampling method applied on the two particle sets, while, finally, in Section 3.4.8 the method for producing a lane hypothesis from the particle set is shown.

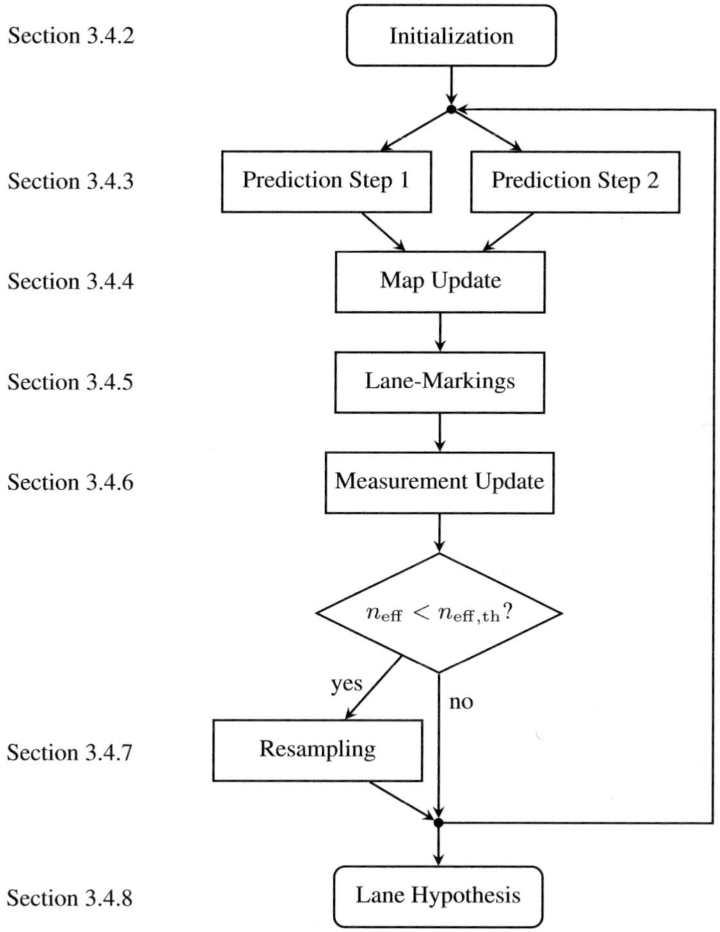

Figure 3.8: Overview of used particle filter algorithm and corresponding sections in this chapter.

3.4.1 System State

The system state chosen for the particle filter is the 2D-pose, i.e. of the form (RABE et al. 2017)

$$\boldsymbol{x}_k = (x_k, y_k, \theta_k)\,, \tag{3.6}$$

describing the vehicle's position in 2D in a local coordinate frame (x_k, y_k) and its heading (θ_k) for time step $k \in \mathbb{N}$. Within this work, the heading θ_k is generally used as a clockwise measure starting with 0 pointing towards the North or y-axis.

To better handle vertical slopes and fully exploit the data from a 3-axis gyroscope, it would be beneficial to instead estimate the 3D-pose, i.e. also the height and roll and pitch angles. However, this would increase the dimensionality of the state space and require a much higher number of particles for a similar result, leading to higher computational load (THRUN et al. 2005). Hence, the system state is restricted to the 2D-pose and slopes are dealt with by increasing the variance of the noise in the position prediction. The distance traveled on a rather steep road with slope of 15% deviates only by around 1.12% from the distance projected onto a flat ground.

As two sources for yaw rate measurements – ESC and directly attached gyroscope – provide acceptable quality but may still lead to trajectories drifting away from the actual position over time, both measurements are incorporated. To do this, one could use another filter stage before the particle filter. In this approach, the particles are divided into two groups that each use one of the sources for yaw rate measurements, very similar to the interacting multiple model (IMM) algorithm (GENOVESE 2001). It can be easily expanded to incorporate further vehicle models such as a bicycle model or one based on visual odometry when suitable sensor data is available. This could also be understood as expanding the system state $\boldsymbol{x}_k$ to a fourth dimension, the discrete value m_k which vehicle model is currently used by a particle:

$$\boldsymbol{x}_k = (x_k, y_k, \theta_k, m_k) \,. \tag{3.7}$$

However, this understanding can become confusing as the vehicle model state m_k is not varied within the prediction step. Hence, during this work, each particle's used vehicle model is treated separately from the other dimensions of the system state.

To check whether a particle is still positioned on a mapped lanelet and to apply the measurement update based on detected lane-markings, the distances and orientation relative to the current lanelet are needed. Therefore, the system state is actually not stored in a local coordinate frame but in lanelet relative coordinates as visualized in Figure 3.9. The currently used lanelet and segments of the left and right polyline boundary are stored together with the lateral distances to the boundaries, the longitudinal offsets to the beginning of the current segment and the angle to the boundary. These can be easily transformed into local or global coordinates.

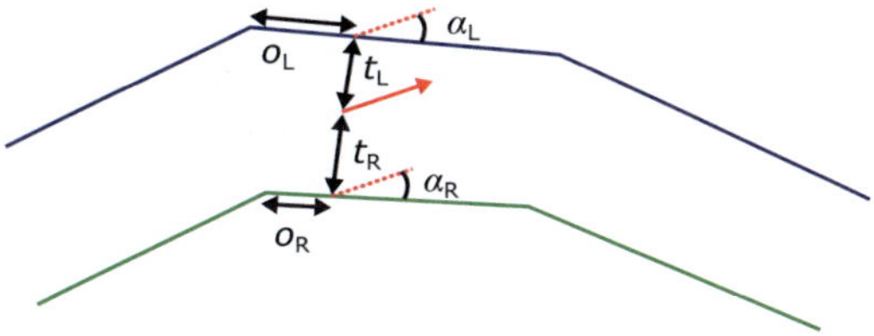

Figure 3.9: Stored particle coordinates for particle shown in red on lanelet with left and right boundaries in blue and green, respectively: Lateral distances t_L, t_R, longitudinal offsets o_L, o_R, and angles α_L, α_R (RABE et al. 2016b).

3.4.2 Initialization

The best guess for the initial pose is provided by GPS. To account for the latency of the GPS pose estimate, this position is predicted using a history of speed and yaw rate measurements and the vehicle model used in the prediction step. The used GPS receiver exhibits only small noise around the actual position but can in some occasions exhibit a bias of more than 20 m. This could be mitigated by first estimating the bias by map-matching or observing the behavior of the position estimate over a longer time period. However, this would mean that the system requires some time and distance traveled before initializing the localization. Therefore, the initial distribution $p\left(\boldsymbol{x}_0|m\right)$ is assumed as a circular uniform distribution with radius of 25 m around the position estimate as proposed in CHAUSSE et al. 2005. This also avoids a significant preference of a single lane over others at initialization as would happen when assuming a Gaussian distribution. The lanelet map m is taken into account by discarding any particle that would be initialized off-road, i.e. not on any lanelet, immediately and generating a new particle. The initial orientation follows a Gaussian distribution around the GPS course estimate. To allow for two particle sets with different prediction steps, each particle is randomly assigned to one of these sets.

3.4.3 Prediction

In the prediction step, the intermediate belief

$$\overline{bel}\left(\boldsymbol{x}_k\right) = p\left(\boldsymbol{x}_k | \boldsymbol{z}_{1:k-1}, \boldsymbol{u}_{1:k}\right) \tag{3.8}$$

before the incorporation of the current measurement $\boldsymbol{z}_k$ is estimated based on the previous state $\boldsymbol{x}_{k-1}$ and the control input $\boldsymbol{u}_k$ by sampling from

$$p\left(\boldsymbol{x}_k | \boldsymbol{x}_{k-1}, \boldsymbol{u}_k\right) \tag{3.9}$$

(THRUN et al. 2005). In practice, this transition is performed using the motion model and the control input consists of noisy measurements of the current vehicle speed and yaw rate which are in general received asynchronously. Accordingly, the prediction step is divided into separate position and orientation predictions. The orientation prediction estimates the current vehicle heading from the yaw rate measurement $\dot{\theta}_k$ as

$$\theta_k^{[i]} = \theta_{k-1}^{[i]} - \Delta t_k \cdot \left(\dot{\theta}_k + \nu_k^{[i]}\right) \tag{3.10}$$

with the realization $\nu_k^{[i]}$ of $\nu_k \sim \mathcal{N}\left(0\frac{\circ}{\mathrm{s}}, \sigma_\theta^2\right)$, where the standard deviation σ_θ depends on the used yaw rate source. Depending on which particle set the particle belongs to, either the bias-corrected ESC or directly attached gyroscope yaw rate measurement and the respective time difference Δt_k is used.
The position prediction is the same for both particle sets. Each particle i is moved forward in its heading direction following[1]

$$x_k^{[i]} = x_{k-1}^{[i]} + d_k^{[i]} \cdot \sin\left(\theta_{k-1}^{[i]}\right) \tag{3.11}$$

$$y_k^{[i]} = y_{k-1}^{[i]} + d_k^{[i]} \cdot \cos\left(\theta_{k-1}^{[i]}\right) \tag{3.12}$$

where d_k is the distance estimate

$$d_k^{[i]} = \Delta t_k \cdot \left(v_{\mathrm{c},k} + \eta_k^{[i]}\right) \tag{3.13}$$

[1] Please note that the sine and cosine terms in Equations (3.11) and (3.12) are switched compared to Equation (2.1) due to the definition of the heading θ_k as the clockwise angle with $\theta_k = 0$ oriented toward the y-axis in this work as opposed to the definition of $\theta_k = 0$ oriented toward the x-axis and increasing counter-clockwise in ESKANDARIAN 2012.

where $v_{c,k}$ is the speed measurement at time step k corrected as described in Equation (3.5) and $\eta_k^{[i]}$ is the realization of a Gaussian random variable following

$$\eta_k \propto \begin{cases} \mathcal{N}\left(0\frac{\mathrm{m}}{\mathrm{s}}, 0.01^2 v_{c,k}^2\right), & v_{c,k} \geq 10\frac{\mathrm{m}}{\mathrm{s}} \\ \mathcal{N}\left(0\frac{\mathrm{m}}{\mathrm{s}}, 0.1^2 \frac{\mathrm{m}^2}{\mathrm{s}^2}\right), & v_{c,k} < 10\frac{\mathrm{m}}{\mathrm{s}}. \end{cases} \tag{3.14}$$

Despite Section 3.2.2 showing that the deviation in estimated distance traveled is not increasing for higher speeds, a noise component proportional to the current speed measurement is chosen for speeds above $10\frac{\mathrm{m}}{\mathrm{s}}$. This is due to experience showing larger deviations between the position estimated by the particle filter and the actual position for higher speeds, likely occurring because of suboptimal sampling.

3.4.4 Map Update

An environment map m can be incorporated into the belief

$$\overline{bel}_m\left(\boldsymbol{x}_k\right) = p\left(\boldsymbol{x}_k | \boldsymbol{z}_{1:k-1}, \boldsymbol{u}_{1:k}, m\right) \tag{3.15}$$

by considering the probability $p\left(\boldsymbol{x}_k | m\right)$ in the prediction step as

$$\begin{aligned} p\left(\boldsymbol{x}_k | \boldsymbol{u}_k, \boldsymbol{x}_{k-1}, m\right) &\propto p\left(m | \boldsymbol{u}_k, \boldsymbol{x}_{k-1}, \boldsymbol{x}_k\right) p\left(\boldsymbol{x}_k | \boldsymbol{u}_k, \boldsymbol{x}_{k-1}\right) \\ &\approx p\left(m | \boldsymbol{x}_k\right) p\left(\boldsymbol{x}_k | \boldsymbol{u}_k, \boldsymbol{x}_{k-1}\right) \\ &= \frac{p\left(\boldsymbol{x}_k | m\right) p(m)}{p\left(\boldsymbol{x}_k\right)} p\left(\boldsymbol{x}_k | \boldsymbol{u}_k, \boldsymbol{x}_{k-1}\right) \\ &\propto \frac{p\left(\boldsymbol{x}_k | m\right) p\left(\boldsymbol{x}_k | \boldsymbol{u}_k, \boldsymbol{x}_{k-1}\right)}{p\left(\boldsymbol{x}_k\right)} \end{aligned} \tag{3.16}$$

(THRUN et al. 2005). The approximation

$$p\left(m | \boldsymbol{u}_k, \boldsymbol{x}_{k-1}, \boldsymbol{x}_k\right) \approx p\left(m | \boldsymbol{x}_k\right) \tag{3.17}$$

omits the pose history leading up to $\boldsymbol{x}_k$. This may lead to likelihoods larger than zero for pose estimates that moved through obstacles. Hence, for this approximation to be valid, high-frequency pose updates have to be ensured together with bounded process noise in the prediction step.

It has to be noted that the map m itself is not a random variable and hence probabilities conditioned on m are not to be understood in a mathematical sense but rather as an intuitive description of a model for the environment that allows for vehicles to travel on the road and with an orientation similar to the orientation of the road at that point. As the area besides any lanelets is considered impassable, $p\left(\boldsymbol{x}_k|m\right) = 0$ for all offroad positions.
In practice, this is implemented as follows: After every move prediction, every particle is checked whether it is still within the boundaries of its previous lanelet. In case a particle has left its previous lanelet to the left or right, it is assigned to a respective neighbor lanelet if one exists. Otherwise, the particle is discarded. When a particle leaves its previous lanelet towards the beginning or end, the same procedure is performed. However, a lanelet may have multiple precursors or successors: in this case, the particle is multiplied and travels on on each of the precursing or succeeding lanelets. The weights of the copies are not changed as most of them will soon leave their respective lanelets and be discarded.
Consider the case where a lane splits into two successors, A and B, that continue into different roads: Say hypothesis A is correct and the odometry measurements describe a vehicle movement that agrees with the shape of lanelet A – and, possibly, its respective successors. On the other hand, hypothesis B describes a noticeable different shape. Then, particles following hypothesis B will soon reach the boundary of their respective lanelet. Then, for these particles a suitable neighboring or succeeding will be searched. However, in most cases, this lanelet will not have a neighbor on the respective side as hypothesis A has just split away on that side. Hence, most of the particles having followed hypothesis B will leave their lanelet and be discarded. The particles following hypothesis A, however, will still have a weight that reflects its importance relative to possible other hypotheses C, D, etc.
In case neither hypothesis A nor B are correct, it is possible in a similar way that the motion model agrees with the shape of one of the hypotheses' lanelets. Then again, the particles following this respective hypothesis will continue to have a weight that describes its importance relative to the correct hypothesis. Alternatively, it is of course possible that particles on both hypotheses will soon be discarded. It is not possible that both hypotheses are correct and very unlikely that both contain lanelets that describe the future vehicle movement very well.

This procedure will lead to a situation where the estimate for the absolute position is false for a short period of time. However, it ensures the correct ratios of lane probabilities after the next characteristic piece of vehicle trajectory for the used model where all particles are always assigned to a certain lanelet.
Besides the position components x_k, y_k of the system state $\boldsymbol{x}_k$, the map also provides information on the orientation θ_k. When a vehicle is traveling on the road, it is likely that it is oriented similar to the current road segment. As stated above, it is crucial to perform weight updates based on the map often enough to avoid particles moving through impassable regions. For the yaw rate, the area to avoid moving through is an orientation perpendicular to the current road segment. With the measured yaw rates and bounded noise variables, this will not happen even within a few time instances. Additionally, this assumption is only coarse compared to orientation measurements from lane-marking detection. Due to these reasons, the orientation map update is synchronized with lane-marking detections and only evaluated when no lane-markings are detected. Then, a weight

$$w^{[i]}_{\mathrm{map},\theta} = \max\left\{\cos\left(\theta^{[i]}_{\mathrm{map,L},k} + \theta^{[i]}_{\mathrm{map,R},k}\right), w_{\mathrm{map},\theta,\mathrm{min}}\right\} \tag{3.18}$$

is determined, where $\theta^{[i]}_{\mathrm{map},S,k}$, $S \in \{\mathrm{L}, \mathrm{R}\}$ denotes the orientation difference to the left and right lane-boundary at the current position, respectively. The weight function using the sum of both angles rather than their average may seem unintuitive and not resemble a proper likelihood but has lead to reasonable results in experiments.
If lane-markings are observed, a weight based on the measured and actual orientation to the lane-markings is determined as described in Section 3.4.6.3.

3.4.5 Combined Weight Update and Sampling based on Detected Lane-Markings

Measurements from the lane-marking detection give a good estimate for the position within and the orientation towards the currently used lane. Despite the intuitive assumption that the position within the lane is negligible for determining the currently used lane if lane-changes are reported when the vehicle laterally passes a lane-marking, knowing the position and orientation within the lane allows for noticing lane-changes even by odometry when no lane-markings can be detected. In this section, the spline model used for the polylines is described first. The concept of combined weight update and sampling (CWUS) is motivated and described in Section 3.4.5.2 and extended to a more general version in Section 3.4.5.3.

3.4.5.1 Spline Interpretation of Lane Boundaries

The mapped polyline boundaries of the lanelets do not perfectly represent the actual road boundaries. Therefore, optimized geometric Hermite (OGH) curves as introduced by YONG and CHENG 2004 and applied to lane centerlines by PETRICH et al. 2013 are used to interpolate between mapped points. OGH curves are cubic splines that are mathematically smooth and minimize strain energy.

With two points $\boldsymbol{P}_0$, $\boldsymbol{P}_1$ at the ends of a segment and two tangent vectors $\boldsymbol{V}_0$, $\boldsymbol{V}_1$ at these points, compare Figure 3.10a, an OGH curve can be expressed as

$$\begin{aligned}\boldsymbol{Q}(s) &= (2s+1)(s-1)^2\boldsymbol{P}_0 + (-2s+3)s^2\boldsymbol{P}_1 \\ &\quad + (1-s)^2 s a_0 \boldsymbol{V}_0 + (s-1)s^2 a_1 \boldsymbol{V}_1\end{aligned} \tag{3.19}$$

for $s \in [0,1]$, where

$$a_0 = \frac{6\langle \boldsymbol{P}_1 - \boldsymbol{P}_0, \boldsymbol{V}_0\rangle \|\boldsymbol{V}_1\|^2 - 3\langle \boldsymbol{P}_1 - \boldsymbol{P}_0, \boldsymbol{V}_1\rangle \langle \boldsymbol{V}_0, \boldsymbol{V}_1\rangle}{4\|\boldsymbol{V}_0\|^2\|\boldsymbol{V}_1\|^2 - \langle \boldsymbol{V}_0, \boldsymbol{V}_1\rangle^2}, \tag{3.20}$$

$$a_1 = \frac{3\langle \boldsymbol{P}_1 - \boldsymbol{P}_0, \boldsymbol{V}_0\rangle \langle \boldsymbol{V}_0, \boldsymbol{V}_1\rangle - 6\langle \boldsymbol{P}_1 - \boldsymbol{P}_0, \boldsymbol{V}_1\rangle \|\boldsymbol{V}_0\|^2}{\langle \boldsymbol{V}_0, \boldsymbol{V}_1\rangle^2 - 4\|\boldsymbol{V}_0\|^2\|\boldsymbol{V}_1\|^2} \tag{3.21}$$

(YONG and CHENG 2004). In the middle of lanelets, the tangent vectors can be defined as

$$\boldsymbol{V}_0 = \boldsymbol{P}_1 - \boldsymbol{P}_{-1}, \tag{3.22}$$

$$\boldsymbol{V}_1 = \boldsymbol{P}_2 - \boldsymbol{P}_0 \tag{3.23}$$

for points $\boldsymbol{P}_{-1}$ and $\boldsymbol{P}_2$ before and after the current segment, respectively. At lanelet beginnings or ends, the respective neighboring points $\boldsymbol{P}_{-1}$, $\boldsymbol{P}_2$ are the mean coordinates of the second last points of the previous and second points of the following lanelets, respectively. In case of no previous or following lanelets, they are defined in direction of the last segment as

$$\boldsymbol{P}_{-1} = 2\boldsymbol{P}_0 - \boldsymbol{P}_1, \tag{3.24}$$

$$\boldsymbol{P}_2 = 2\boldsymbol{P}_1 - \boldsymbol{P}_0. \tag{3.25}$$

A resulting OGH curve it depicted in Figure 3.10b.

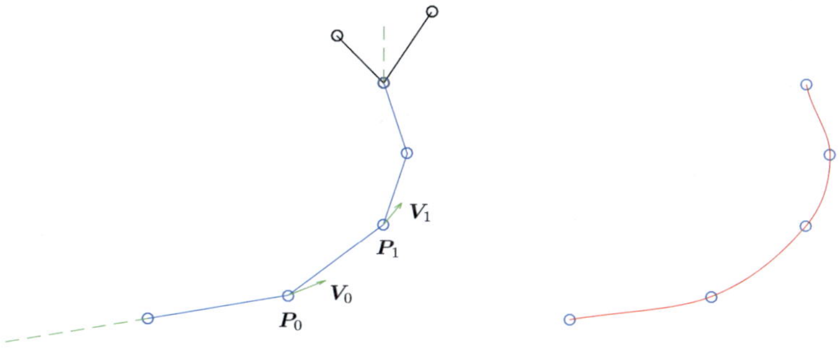

(a) Map input.

(b) OGH spline.

Figure 3.10: Map input as polyline in blue and corresponding OGH spline for example map segment. Points $\boldsymbol{P}_0, \boldsymbol{P}_1$ and tangent vectors $\boldsymbol{V}_0, \boldsymbol{V}_1$ for the second line segment are annotated. First segments of two succeeding lanelet boundaries shown in black, assumed first and last segment tangents in dashed green.

To determine the distance and angle of a particle to this curve, first the parameter s that minimizes the distance

$$\boldsymbol{r}(s) = \boldsymbol{Q}(s) - \boldsymbol{x}_k^{[i]} \tag{3.26}$$

from the curve to the particle in a least-squares sense is determined using the Gauss-Newton algorithm. When using the relative longitudinal offset of a particle within a segment – compare Figure 3.9 – as initial guess

$$s_0 = \frac{o_S}{\|\boldsymbol{P}_1 - \boldsymbol{P}_0\|}, S \in \{\mathrm{L}, \mathrm{R}\}, \tag{3.27}$$

the estimated longitudinal offset

$$s_i = s_{i-1} - \frac{\langle \boldsymbol{Q}'(s_{i-1}), \boldsymbol{r}(s_{i-1}) \rangle}{\|\boldsymbol{Q}'(s_{i-1})\|^2} \tag{3.28}$$

exhibits only minor changes $|s_i - s_{i-1}| < 10^{-3}$ in many evaluated cases. Hence, within this work, only one step of this iterative algorithm is performed. With this, the current distance to the lane boundary is the distance to the point $\boldsymbol{Q}(s_1)$ and the yaw difference is the difference between the direction of the tangent $\boldsymbol{Q}'(s_1)$ and the global heading of the particle.

3.4.5.2 Combined Weight Update and Sampling

In a straight-forward manner, the particles would now be weighted according to the difference between their observed distance and yaw to the lane boundaries and the respective values determined from their position on the lanelet map. But as the likelihood is a lot tighter than the intermediate belief at the first lane-marking observation after initialization, this can quickly lead to unwanted effects: In a small simulation, a set of 50 randomly drawn particles is initialized on three lanes as shown in the example realization in Figure 3.11a. Process noise from the prediction step is emulated by having the particles move slightly around their original position. All sensor input is ignored except for the distance to the lane-markings – without loss of generality, it provides a constant distance estimate of 2 m to both the left and right marking, leading to the likelihood shown in Figure 3.11b. After 10 filter steps of "prediction", measurement update, and resampling, the particle distribution may look like the one in Figure 3.11c. Only few particles lie close to the likelihood's peaks and are assigned a high weight. Over time, clusters develop around a few particles that have been initialized close to these maxima. As these surviving particles are in general not equally distributed over several lanes and spread over them differently due to different initial heading angles, the particle set does not represent the distribution that we expect from the sensor measurements very well anymore: The particle sets on the center and right lane are concentrated in four groups – despite no sensor having given any information about the longitudinal position on the road. This sample impoverishment has also been noted by MILSTEIN et al. 2002.

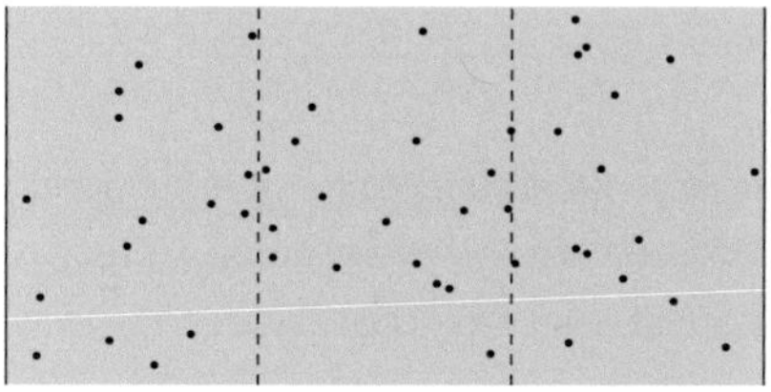

(a) Initial particle distribution.

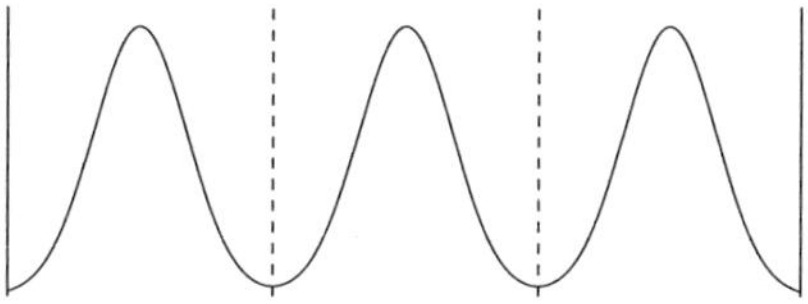

(b) Likelihood.

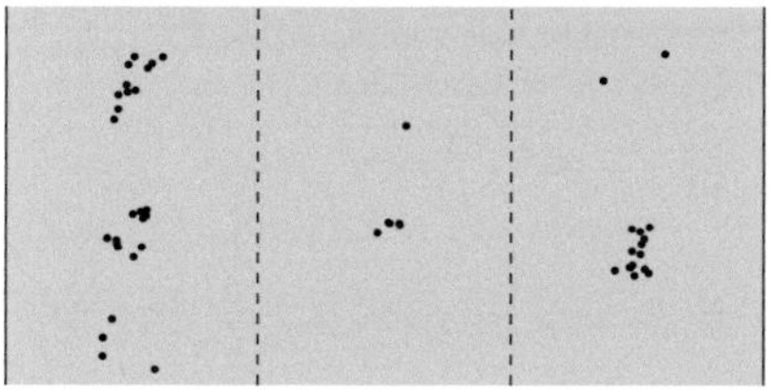

(c) After 10 steps of prediction, weighting, and resampling.

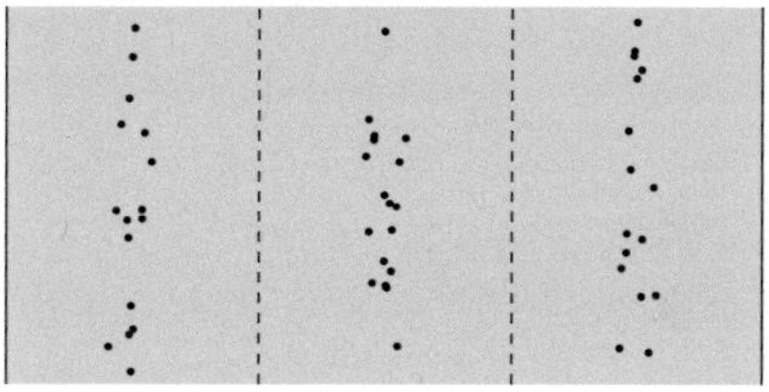

(d) After 10 steps of prediction, combined weight update and sampling, and resampling.

Figure 3.11: Exemplary particle distributions on a three-lane road segment at initialization (a) and after 10 filter steps (c, d). Exemplary Gaussian likelihood for lane-marking detection with standard deviation $\sigma_{\mathrm{m},k} = 0.5\,\mathrm{m}$ for lane-width $w_\mathrm{l} = 4\,\mathrm{m}$ (b) (RABE et al. 2017). See text for more detailed description.

Furthermore, the lane sum weights are supposed to indicate the lane probabilities. After the last step of resampling – i.e. all particles have the same weight and the number of particles indicates the lane probability as well –, there are now 27, 6, and 17 particles on the respective lanes. Again, no sensor observation should have discriminated the center lane against the left lane.

As this unwanted effect is strongest in sparse particle sets and tight likelihoods, there are intuitive solutions against it: By increasing the number of particles, the set becomes more dense and the effect is reduced. However, this also increases the computational load. Alternatively, one could widen up the likelihood and have more particles partially agree with it. This would mainly reduce how fast the effect strikes and at the same time discard valuable information. Similarly, increasing the process noise fights the symptom of clustering as the particles fill up the empty spaces more quickly – but again discards information from odometry.

To make the best out of the computational resources and the available sensor data, this section describes an alternative approach to the problem following the ideas of the Kalman filter (KALMAN 1960) and the *"likelihood" particle filter* mentioned in ARULAMPALAM et al. 2002. Using a combined weight update and sampling step, one can achieve a particle distribution as shown in Figure 3.11d after 10 filter steps. One can easily notice that the longitudinal distribution is much less clustered than before and there are 16, 18, and 16 particles, respectively, on the three lanes – just as after initialization.

In the approach by Arulampalam et al., the particles are not sampled from the intermediate belief and then weighted according to the likelihood but instead sampled from the likelihood and weighted according to the intermediate belief to estimate the product of both. When instead assuming the lateral component of intermediate beliefs and likelihoods to be Gaussian, their product can be calculated – resembling a Kalman filter approach – and then the new set of particles can be sampled from the product distribution.

First, the mean $\mu_{\mathrm{p},s,k}$ and standard deviation $\sigma_{\mathrm{p},s,k}$ of the particles constituting the intermediate belief are estimated separately for each segment s of both left and right boundaries of each currently used lanelet using a maximum likelihood estimator

$$\hat{\mu}_{\mathrm{p},s,k} = \frac{\sum_{i=1}^{n_s} w_k^{[i]} x_k^{[i]}}{\sum_{i=1}^{n_s} w_k^{[i]}} \tag{3.29}$$

and

$$\hat{\sigma}^2_{\mathrm{p},s,k} = \frac{\sum_{i=1}^{n_s} w_k^{[i]} \left(x_k^{[i]} - \hat{\mu}_{\mathrm{p},s,k} \right)^2}{\left(\sum_{i=1}^{n_s} w_k^{[i]} \right) - 1} \tag{3.30}$$

for n_s particles in segment s. At forming lanes, on rural roads without center lane-marking, or in situations with bad road marking, the detected lane might not coincide with the boundary of the currently used lanelet. Therefore, also the distance and orientation to the boundaries of possible neighboring, oncoming, and – at splits – other lanelets are considered by finding the boundary segment s for each particle i that minimizes the deviation between position on the map and detected distance. When both the left and right lane-marking are detected, boundary segments are chosen that agree well for both sides – otherwise situations may occur where the determined corresponding boundary is the same for both sides.

The means $\mu_{\mathrm{m},k}$ for the likelihoods are the observed distances to the left and right lane-markings, their standard deviations $\sigma_{\mathrm{m},k}$ are fixed beforehand. The overall distribution for the respective boundary segment s is then a Gaussian $\mathcal{N}\left(\mu_{\mathrm{c},s,k}, \sigma^2_{\mathrm{c},s,k}\right)$ with (RABE et al. 2017)

$$\mu_{\mathrm{c},s,k} = \frac{\mu_{\mathrm{p},s,k}\sigma^2_{\mathrm{m},k} + \mu_{\mathrm{m},k}\sigma^2_{\mathrm{p},s,k}}{\sigma^2_{\mathrm{p},s,k} + \sigma^2_{\mathrm{m},k}} \tag{3.31}$$

and

$$\sigma_{\mathrm{c},s,k} = \frac{\sigma_{\mathrm{p},s,k}\sigma_{\mathrm{m},k}}{\sqrt{\sigma^2_{\mathrm{p},s,k} + \sigma^2_{\mathrm{m},k}}}. \tag{3.32}$$

Sampling from the product distribution can be performed by shifting each particle i laterally by (RABE and STILLER 2017)

$$\Delta x_k^{[i]} = \mu_{\mathrm{c},s,k} - \frac{\sigma_{\mathrm{c},s,k}}{\sigma_{\mathrm{p},s,k}}\mu_{\mathrm{p},s,k} + \left(\frac{\sigma_{\mathrm{c},s,k}}{\sigma_{\mathrm{p},s,k}} - 1\right) x_k^{[i]}. \tag{3.33}$$

This shift describes how a sample from the intermediate belief distribution following $\mathcal{N}(\mu_{\mathrm{p},s,k}, \sigma^2_{\mathrm{p},s,k})$ should be moved in space to become a sample from $\mathcal{N}(\mu_{\mathrm{c},s,k}, \sigma^2_{\mathrm{c},s,k})$. It is motivated as depicted in Fig. 3.12. First, a sample x_k from the intermediate distribution is taken and transformed to a sample of the standard normal distribution:

$$x_{\mathrm{s}} = \frac{x_k - \mu_{\mathrm{p},s,k}}{\sigma_{\mathrm{p},s,k}}. \tag{3.34}$$

This sample of the standard normal distribution can then be transformed to a sample of the new distribution as

$$x_k' = \sigma_{\mathrm{c},s,k}x_{\mathrm{s}} + \mu_{\mathrm{c},s,k}. \tag{3.35}$$

The distance between both samples is

$$\begin{aligned}
x'_k - x_k &= \sigma_{\mathrm{c},s,k} x_\mathrm{s} + \mu_{\mathrm{c},s,k} - x_k \\
&= \sigma_{\mathrm{c},s,k} \frac{x_k - \mu_{\mathrm{p},s,k}}{\sigma_{\mathrm{p},s,k}} + \mu_{\mathrm{c},s,k} - x_k \\
&= \frac{\sigma_{\mathrm{c},s,k}}{\sigma_{\mathrm{p},s,k}} \left(x_k - \mu_{\mathrm{p},s,k} \right) + \mu_{\mathrm{c},s,k} - x_k \\
&= \frac{\sigma_{\mathrm{c},s,k}}{\sigma_{\mathrm{p},s,k}} x_k - \frac{\sigma_{\mathrm{c},s,k}}{\sigma_{\mathrm{p},s,k}} \mu_{\mathrm{p},s,k} + \mu_{\mathrm{c},s,k} - x_k \\
&= \mu_{\mathrm{c},s,k} - \frac{\sigma_{\mathrm{c},s,k}}{\sigma_{\mathrm{p},s,k}} \mu_{\mathrm{p},s,k} + \left(\frac{\sigma_{\mathrm{c},s,k}}{\sigma_{\mathrm{p},s,k}} - 1 \right) x_k .
\end{aligned} \tag{3.36}$$

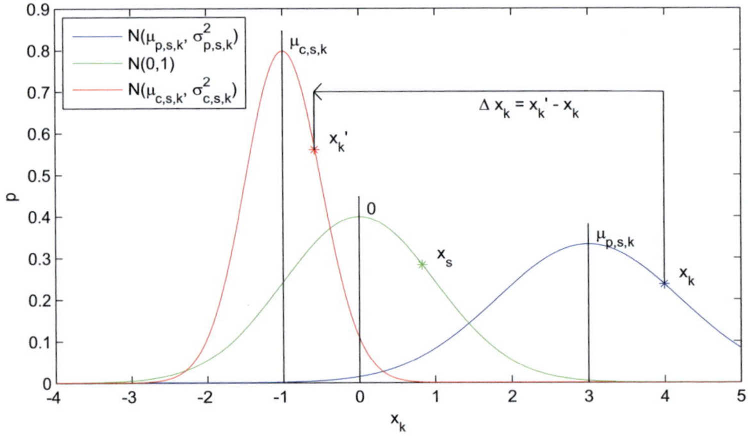

Figure 3.12: Transformation of a sample from one normal distribution $\mathcal{N}(\mu_{\mathrm{p},s,k}, \sigma^2_{\mathrm{p},s,k})$ to another $\mathcal{N}(\mu_{\mathrm{c},s,k}, \sigma^2_{\mathrm{c},s,k})$ via the standard normal distribution.

By applying the lateral shift described in Equation (3.33), the set of particles approximates the product of the intermediate belief and likelihood for the lateral position component by the particle positions without weights being assigned. This approach is not only applied directly after initialization, where the assumed GPS distribution is much wider than the one of the lane-marking detection, but also helps to keep consistency with the map in case of map errors such as inaccurately digitized lane boundaries. While the dead reckoning might follow the true path of the vehicle, the goal is the localization relative to the map and with this approach, the dead reckoning trajectory is corrected to comply with the map without widening up the intermediate beliefs with huge process noise in the prediction step.

3.4.5.3 Robust Combined Weight Update and Sampling

The approach described in the previous section is already very useful in terms of performance and reproducibility of lane-precise localization. However, the assumptions of normal distributed intermediate belief and likelihood are not perfect: As described in Section 3.4.2, the initial particle distribution $p(\boldsymbol{x}_0)$ is uniformly distributed in the position dimensions. Hence, the intermediate belief $\overline{bel}(\boldsymbol{x}_0)$ will be far from a normal distribution at the first lane-marking observation. Additionally, the camera might detect other lines in its image than the actual lane-markings, the lanelet boundaries in the map may not describe the lane-markings well or the actually corresponding marking may not be found. To be more robust against such outages, the likelihood is typically not chosen as a pure normal distribution, but might be a lower-bounded normal distribution – corresponding to an upper bounded influence function Ψ – or a mixture likelihood consisting of a normal and a uniform part. Therefore, the approach is extended to mixture distributions for both the intermediate belief and the likelihood. To achieve this, it is necessary to first estimate the intermediate belief $\overline{bel}(\boldsymbol{x}_k)$ and then transform the particles to fit the product distribution.

Estimating the Intermediate Belief When defining the parameter vector

$$\Theta = \{\theta_1, \theta_2\} = \{\{\pi_1, a_1, b_1\}, \{\pi_2, \mu_1, \sigma_1\}\} \tag{3.37}$$

with $\pi_2 = 1 - \pi_1$ and shorthands $\mathcal{U}_i$ and $\mathcal{N}_i$ for the uniform distribution

$$\mathcal{U}_i = \mathcal{U}(x; a_i, b_i) = \begin{cases} \frac{1}{b_i - a_i}, & x \in [a_i, b_i] \\ 0, & \text{else,} \end{cases} \tag{3.38}$$

and the normal distribution

$$\mathcal{N}_i = \mathcal{N}\left(x; \mu_i, \sigma_i\right) = \frac{1}{\sqrt{2\pi}\sigma_i} \exp\left(-\frac{\left(x-\mu_i\right)^2}{2\sigma_i^2}\right), \tag{3.39}$$

the problem of estimating the mixture intermediate belief can be written as estimating a distribution of the form (RABE and STILLER 2017)

$$\begin{aligned} g\left(x; W, \Theta\right) &= \pi_1 f_1\left(x; W, \theta_1\right) + \pi_2 f_2\left(x; W, \theta_2\right) \\ &= \pi_1 \mathcal{U}_1 + \left(1-\pi_1\right)\mathcal{N}_1 \end{aligned} \tag{3.40}$$

with particle weights $W = \left\{w^{[1]}, \ldots, w^{[N]}\right\}$. To handle each particle according to which part of the distribution it belongs to, the probability (RABE and STILLER 2017)

$$\phi_{i,j} = p\left(z^{[i]} = j \middle| x^{[i]}; w^{[i]}, \Theta\right) \tag{3.41}$$

that a particle i belongs to the component $f_j, j \in \{1, 2\}$ has to be estimated as well. This can be done using an iterative expectation maximization (EM) algorithm to maximize the expected complete data log likelihood (CORETTO and HENNIG 2011; GEBRU et al. 2016)

$$\hat{\Theta} = \arg\max_{\Theta} E\left[\ln\left(g\left(X|W, \Theta\right)\right)\right]. \tag{3.42}$$

The expectation can be written as (RABE and STILLER 2017)

$$\begin{aligned} Q\left(\Theta, \Theta^{(t)}\right) &= E\left[\ln\left(P\left(X, Z|W, \Theta\right)\right)\right] \\ &= \sum_{i=1}^{N} \ln\left(\sum_{j=1}^{2} \pi_j f_j\left(x; \theta_j^{(t)}, w^{[i]}\right)\right) \end{aligned} \tag{3.43}$$

for step t of the algorithm with

$$f_1(x; \theta_1, w^{[i]}) = \left(\mathcal{U}\left(x; a_1, b_1\right)\right)^{w^{[i]}} \tag{3.44}$$

and

$$f_2(x; \theta_2, w^{[i]}) = \left(\mathcal{N}\left(x; \mu_1, \sigma_1\right)\right)^{w^{[i]}}. \tag{3.45}$$

with the weight $w^{[i]}$ of particle i. Any $w^{[i]} \neq 1$ corresponds to a $w^{[i]}$-fold observation of the respective sample or particle. This is intuitive for integer

$w^{[i]}$ but can be easily extended to real numbered weights (GEBRU et al. 2016). In the next step, GEBRU et al. replace $\left(\mathcal{N}\left(\mu,\sigma^2\right)\right)^{w^{[i]}}$ by $\mathcal{N}\left(\mu,\frac{\sigma^2}{w^{[i]}}\right)$, which is proportional to $\left(\mathcal{N}\left(\mu,\sigma^2\right)\right)^{w^{[i]}}$. While this scaling may be irrelevant in the Gaussian-only mixture GEBRU et al. are dealing with, it has to be taken care of in the uniform/normal mixture case. Hence, it seems more intuitive to use $\left(\mathcal{N}\left(\mu,\sigma^2\right)\right)^{w^{[i]}}$ and carry $w^{[i]}$ through the derivation, leading to (RABE and STILLER 2017)

$$\begin{aligned}
Q\left(\Theta,\Theta^{(t)}\right) &= \sum_{i=1}^{N}\sum_{j=1}^{2}\left(\phi_{i,j}^{(t)}w^{[i]}\ln\left(\pi_j f_j\right)\right)\\
&= \sum_{i=1}^{N}\left(\phi_{i,1}^{(t)}w^{[i]}\ln\left(\pi_1 f_1\right)+\phi_{i,2}^{(t)}w^{[i]}\ln\left(\pi_2 f_2\right)\right)\\
&= \sum_{i=1}^{N}\left(\phi_{i,1}^{(t)}w^{[i]}\left(\ln\pi_1+\ln\frac{\mathbf{1}_{[a_1,b_1]}}{b_1-a_1}\right)\right)\\
&\quad+\sum_{i=1}^{N}\left(\phi_{i,2}^{(t)}w^{[i]}\left(\ln\frac{1-\pi_1}{\sqrt{2\pi}\sigma_1}-\frac{\left(x_i-\mu_1\right)^2}{2\sigma_1^2}\right)\right)
\end{aligned} \tag{3.46}$$

with $f_j = f_j\left(x;\theta_j^{(t)},1\right)$ for better readability. Following CORETTO and HENNIG 2011, one may choose $a_1 = \min_i x^{[i]}$ and $b_1 = \max_i x^{[i]}$. The other parameters can be found through derivation: As

$$\frac{\partial Q\left(\Theta,\Theta^{(t)}\right)}{\partial\mu_1} = \sum_{i=1}^{N}\left(\phi_{i,2}^{(t)}w^{[i]}\frac{x^{[i]}-\mu_1}{\sigma_1^2}\right)\overset{!}{=}0, \tag{3.47}$$

$$\frac{\partial Q\left(\Theta,\Theta^{(t)}\right)}{\partial\sigma_1} = \sum_{i=1}^{N}\left(\phi_{i,2}^{(t)}w^{[i]}\left(-\frac{1}{\sigma_1}+\frac{x^{[i]}-\mu_1}{\sigma_1^3}\right)\right)\overset{!}{=}0, \tag{3.48}$$

and

$$\frac{\partial Q\left(\Theta,\Theta^{(t)}\right)}{\partial\pi_1} = \sum_{i=1}^{N}\left(\phi_{i,1}^{(t)}\frac{1}{\pi_1}+\phi_{i,2}^{(t)}\frac{1}{\pi_1-1}\right)\overset{!}{=}0, \tag{3.49}$$

we find the parameters for the next iteration as (RABE and STILLER 2017)

$$\mu_1^{(t+1)} = \frac{\sum_{i=1}^{N}\phi_{i,2}^{(t+1)}w^{[i]}x^{[i]}}{\sum_{i=1}^{N}\phi_{i,2}^{(t+1)}w^{[i]}}, \tag{3.50}$$

$$\sigma_1^{(t+1)} = \sqrt{\frac{\sum_{i=1}^{N} \phi_{i,2}^{(t+1)} w^{[i]} \left(x^{[i]} - \mu_1^{(t+1)}\right)^2}{\sum_{i=1}^{N} \phi_{i,2}^{(t+1)} w^{[i]}}}, \tag{3.51}$$

and

$$\pi_1^{(t+1)} = \frac{\sum_{i=1}^{N} \phi_{i,1}^{(t+1)} w^{[i]}}{\sum_{i=1}^{N} w^{[i]} \left(\phi_{i,1}^{(t+1)} + \phi_{i,2}^{(t+1)}\right)} = \frac{\sum_{i=1}^{N} \phi_{i,1}^{(t+1)} w^{[i]}}{\sum_{i=1}^{N} w^{[i]}}, \tag{3.52}$$

with

$$\phi_{i,j}^{(t+1)} = \frac{\pi_j^{(t)} f_j \left(x^{[i]}; w^{[i]}, \theta_j^{(t)}\right)}{g\left(x; W, \Theta^{(t)}\right)}. \tag{3.53}$$

Iterating can be stopped when the change in the parameters becomes small.

Sample Transformation The intermediate belief estimated this way can be used to extend the combined weight update and sampling step to mixture distributions. As the parameters of the likelihood are self-defined except for the mean of the normal part, one can calculate the resulting posterior as (RABE and STILLER 2017)

$$\begin{aligned} &(\pi_1 \mathcal{U}_1 + (1-\pi_1)\mathcal{N}_1) \cdot (\pi_\mathrm{m} \mathcal{U}_\mathrm{m} + (1-\pi_\mathrm{m})\mathcal{N}_\mathrm{m}) = \\ &\pi_1 \pi_\mathrm{m} \mathcal{U}_1 \mathcal{U}_\mathrm{m} + \pi_1 (1-\pi_\mathrm{m}) \mathcal{U}_1 \mathcal{N}_\mathrm{m} + \\ &(1-\pi_1) \pi_\mathrm{m} \mathcal{U}_\mathrm{m} \mathcal{N}_1 + (1-\pi_1)(1-\pi_\mathrm{m}) \mathcal{N}_\mathrm{c} \end{aligned} \tag{3.54}$$

with $\mathcal{U}_i$ and $\mathcal{N}_i$ as in Equations (3.38) and (3.39), respectively, and μ_c and σ_c as in Equations (3.31) and (3.32). $\mathcal{U}_\mathrm{m} = \mathbf{1}_{[a_\mathrm{m}, b_\mathrm{m}]} (b_\mathrm{m} - a_\mathrm{m})^{-1}$ describes the uniform component of the mixture likelihood, where the boundaries a_m and b_m are chosen as the respective distances to the left and right boundaries of the road.

Each particle i can now be transformed to be a sample of the product distribution. This transformation depends on whether it belonged to the uniform or normal part of the intermediate belief and on which part of the likelihood is applied. This can be statistically modeled with realizations of two independent uniform random numbers $r_i, q_i \in [0, 1]$ as depicted in Figure 3.13.

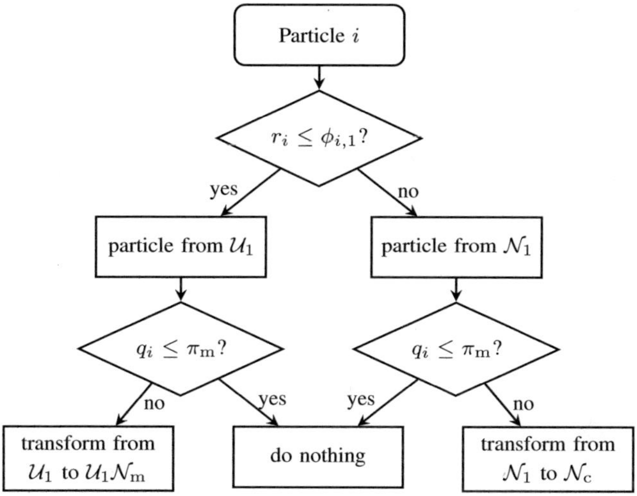

Figure 3.13: Decision tree for combined weight update and sampling with mixture intermediate belief and likelihood, following (RABE and STILLER 2017).

The transformation from $\mathcal{N}_1$ to $\mathcal{N}_\mathrm{c}$ is performed as in Equation (3.33). To transform a sample from the uniform $\mathcal{U}_1$ to the cut off Gaussian $\mathcal{U}_1\mathcal{N}_\mathrm{m}$, one can go via the standard distributions $\mathcal{U}(x; 0, 1)$ and $\mathcal{N}(x; 0, 1)$ (RABE and STILLER 2017). For this purpose, one normalizes

$$x^{[i](1)} = \frac{x^{[i]} - a_1}{b_1 - a_1}, \tag{3.55}$$

then transforms from the uniform to the normal distribution as

$$x^{[i](2)} = \sqrt{2} \cdot \operatorname{erf}^{-1}\left(2x^{[i](1)} - 1\right), \tag{3.56}$$

and finally

$$x^{[i](3)} = \sigma_\mathrm{m} x^{[i](2)} + \mu_\mathrm{m}. \tag{3.57}$$

Like this, the samples $x^{[i]} = a_1$ and $x^{[j]} = b_1$ are transformed to $x^{[i](3)} = -\infty$ and $x^{[j](3)} = \infty$, respectively, which obviously cannot be samples of the cut off Gaussian $\mathcal{U}_1\mathcal{N}_\mathrm{m}$. This can also happen to other samples. Therefore, all resulting samples with $x^{[i](3)} > b_1$ or $x^{[i](3)} < a_1$ are replaced by random samples from $\mathcal{U}_1\mathcal{N}_\mathrm{m}$.

Sample Transformation for Correlated Outliers Errors in the lane-marking detection may stem from misdetections, where e.g. shadows, bitumen lines, guardrails, or other linear structures are detected rather than the actual lane-marking or road boundary. Due to this and the tracking performed within the lane-marking detection, misdetections occur mostly not individually but in batches. Hence, approaching erroneous measurements by randomly assigning a particle to either component of the mixture distribution at each measurement may lead to further errors in the localization.

With this in mind, the particle state can be extended by two boolean variables that define whether lane-markings on the left and right side can currently be trusted. To have particles in the *not reliable* state for more than one time step, this can be modeled as a Markov chain with the transition probabilities p_{12} and p_{21} chosen according to the desired stationary behavior, as seen in Figure 3.14.

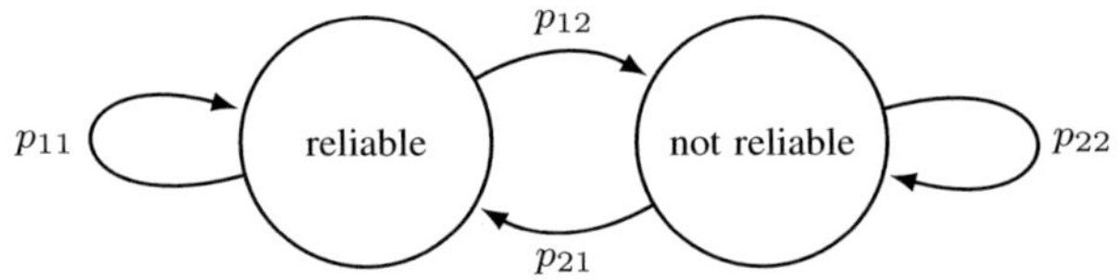

Figure 3.14: Markov chain with states whether lane-marking is reliable or not.

Given a desired stationary *not reliable* probability $p_2 = \pi_{\mathrm{m}}$ and one transition probability, e.g. p_{21}, one can determine the other transition probabilities as

$$\begin{aligned} p_{11} &= \frac{1 - p_2 - (1 - p_{22})\, p_2}{1 - p_2}, \\ p_{12} &= 1 - p_{11}, \\ p_{22} &= 1 - p_{21}. \end{aligned} \tag{3.58}$$

The given transition probability p_{21} can be chosen such that a desired average length of stay in *not reliable* state is achieved. These states are updated in every resampling step using a random number from a uniform distribution $\mathcal{U}(0, 1)$. Now, the particle's state determines whether it is assumed to be part of the uniform or Gaussian part of the likelihood and the lower decisions in Figure 3.13, "$q_i \leq \pi_{\mathrm{m}}$?", are replaced by *"particle in not reliable state?"*.

3.4.6 Measurement Update

In the measurement update step, the weight

$$w_k^{[i]} \propto p\left(\boldsymbol{z}_{1:k} \middle| \boldsymbol{x}_k^{[i]}\right) \tag{3.59}$$

is calculated using the likelihood of incoming measurements $\boldsymbol{z}_k$ given the current particle state. This leads to the set of weighted particles approximating the posterior belief

$$bel = p\left(\boldsymbol{x}_k | \boldsymbol{z}_{1:k}, \boldsymbol{u}_{1:k}\right). \tag{3.60}$$

For each type of measurement $z_{\mathrm{s},k}$ and particle i, a separate weight $w_{\mathrm{s},k}^{[i]}$ is determined. The new overall weight is determined as the product of the individual weights determined in that time instance k with the previous weight of the particle $w_{k-1}^{[i]}$:

$$w_k^{[i]} = w_{k-1}^{[i]} \cdot \prod_{\mathrm{s}} w_{\mathrm{s},k}^{[i]}. \tag{3.61}$$

As the sensor measurements arrive in general asynchronously, the product is calculated only over the respective weights $w_{\mathrm{s},k}^{[i]}$ in time instance k. In the practical implementation, the particle weights are stored in a linear scale and are normalized to sum to 1 whenever a resampling step is performed. Figure 3.15 gives an overview of the sensor data used in the measurement update step which are shown in more detail within this section. Detected lane-markings are depicted as well as their angle is used for a weight update. The use of the lateral distance to them in a combined weight update and sampling step has been motivated and described in Section 3.4.5.

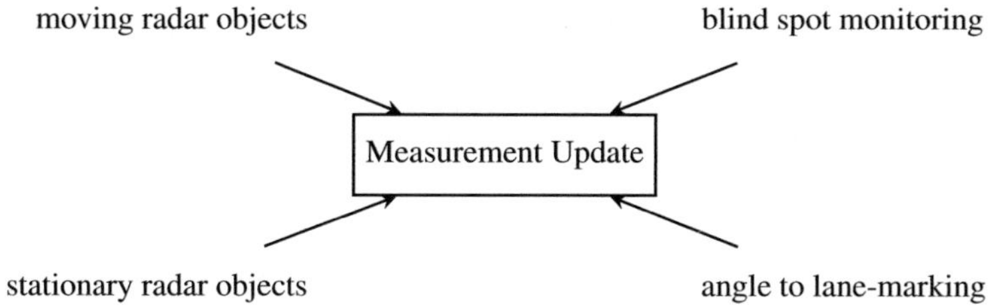

Figure 3.15: Overview of sensor data used in measurement update.

3.4.6.1 Radar Data

Observed radar objects can give a hint where on the roadway the ego-vehicle is located. Intuitively, other moving vehicles should drive on the road while guardrails should be situated next to it. With the relative position of detected and prefiltered objects, every particle can be weighted according to whether an observed object is plausible given its current position. For this purpose, the relative coordinates of an object are projected from each particle and the smallest distance d from the resulting position to the surrounding lanelets is determined, with positions on lanelets denoted with negative distances $d < 0\,\mathrm{m}$. Then, a weight is determined depending on the plausibility of the calculated distance, i.e. (RABE et al. 2017)

$$w_{\mathrm{radar}}^{[i]} = \begin{cases} \exp\left(-\frac{d^2}{2\sigma_{\mathrm{radar}}^2}\right), & d > 0\,\mathrm{m} \\ 1, & \text{else} \end{cases} \tag{3.62}$$

for observed moving cars and trucks and

$$w_{\mathrm{radar}}^{[i]} = \begin{cases} 1, & d > 0\,\mathrm{m} \\ \exp\left(-\frac{d^2}{2\sigma_{\mathrm{radar}}^2}\right), & \text{else} \end{cases} \tag{3.63}$$

for observed guardrails. In both cases, the weight is lower bounded with minimal weights $w_{\mathrm{radar,car,min}}$ and $w_{\mathrm{radar,guardrail,min}}$, respectively.
However, objects might be erroneous in their relative position and classification. Thus, an average contradiction value is determined as

$$r = \sum_{i=1}^{n} \frac{1 - w_{\mathrm{radar}}^{[i]}}{1 - w_{\mathrm{radar},t,\mathrm{min}}}, \tag{3.64}$$

where $w_{\mathrm{radar},t,\mathrm{min}}$ is the minimal weight for the respective object type $t \in \{\mathrm{car}, \mathrm{guardrail}\}$. This summand is chosen such that the importance weights assigned to particles to which the observation complies perfectly add no value to r, those with coarse contradictions add value 1, and those in between are partly counted depending on their distance. If this value exceeds a threshold r_{th}, it is assumed that the measurement is erroneous. In that case, the weights are not applied to the particles.

A series of contradictions can be an indicator that the filter has converged to a wrong lane. Therefore, it is kept track how often an observed car or truck leads to such many contradictions that $r < r_{\text{th}}$. If that happens at least 5 times in one second, a number of new particles is created to provide hypotheses that better fit the current observations. Furthermore, the localization is set unavailable (compare Section 3.4.8) for half a second. Objects classified as guardrails are not used for this partial reinitialization as even after filtering them, they still contain many erroneous measurements.

3.4.6.2 Blind-Spot Monitoring

The idea behind observations from the blind-spot monitoring system is the same as for vehicles detected by the front radar. The difference is that it is only known whether another vehicle is observed in the blind spot and not where exactly it is. Therefore, the weight can only take fixed values for each side. It depends on whether a neighboring lane in the direction of an observed vehicle exists according to the map or not:

$$w_{\text{BSM,L}}^{[i]} = \begin{cases} 1, & \text{lane exists to the left} \\ w_{\text{BSM,min}}, & \text{else} \end{cases} \tag{3.65}$$

for observed vehicles on the left side, and accordingly for the right. If a vehicle is observed on only one side, the overall blind-spot monitoring weight is the weight for the respective side. Otherwise, the overall weight

$$w_{\text{BSM}}^{[i]} = w_{\text{BSM,R}}^{[i]} w_{\text{BSM,L}}^{[i]} \tag{3.66}$$

is used.

As for the radar objects detected by the front radar, the number of contradictions for each observation is counted and the weight is not applied if it exceeds a threshold. Due to the high reliability of the blind-spot monitoring system, a partial reinitialization with new particles is already triggered for two highly contradicting observations in one second.

3.4.6.3 Angle to Lane-Marking

The observed angle $\theta_{\text{marking},k}$ to the lane-marking used in a classical weight update with

$$w_{\text{marking}}^{[i]} = \max\left\{\cos\left(\theta_{\text{marking},k} - \theta_{\text{map},k}^{[i]}\right), w_{\text{marking,min}}\right\} \tag{3.67}$$

with the angle $\theta_{\text{map},k}^{[i]}$ to the tangent to the OGH curve.

3.4.7 Resampling

When importance weights from all measurements have been applied to the particles, the variance in the particle weights has increased. Eventually, the particle set will contain few particles with high weight and many with negligibly low weight, known as particle degeneracy (DOUCET et al. 2000). This problem, however, can be solved by performing a resampling step.

In this approach, resampling is performed whenever the effective sample size (DOUCET et al. 2000)

$$n_{\text{eff}} = \frac{1}{\sum_{i=0}^{n-1}\left(w^{[i]}\right)^2} \tag{3.68}$$

falls below a threshold $n_{\text{th}} = 0.8n$. In the resampling step, new particles are drawn from the original set according to their importance weights $w^{[i]}$. To reduce variance compared to randomly drawing from all particles, a systematic resampling method widely used in literature (DOUC and CAPPÉ 2005; DOUCET and JOHANSEN 2009; HOL et al. 2006; HWANG and SUNG 2013) is used. It should be noted that it is not guaranteed for the general case that this resampling scheme leads to particle sets more consistent with the true probability distribution (DOUC and CAPPÉ 2005).

However, the main goal of the localization system is to figure out the currently used lane and it is crucial to keep the correct hypothesis alive. Therefore, a component of niching known from genetic algorithms (BIENVENÜE et al. 2002) is added where each lanelet is considered as a niche with its respective particle population: The empirical probabilities $\frac{n_l}{n}$ of all used lanelets l after resampling are compared to the estimated probabilities

$$\hat{p}_l = \sum_{\boldsymbol{x}_i \text{ on lanelet } l} w^{[i]} \tag{3.69}$$

before resampling. Three cases can be distinguished:

1. $n_l > \lceil n\hat{p}_l \rceil$: too many new particles; discard $n_l - \lceil n\hat{p}_l \rceil$ random particles on lanelet l from new particle set,
2. $n_l = \lceil n\hat{p}_l \rceil$: new particle number okay, and
3. $n_l < \lceil n\hat{p}_l \rceil$: too few new particles; sample $\lceil n\hat{p}_l \rceil - n_l$ additional particles on lanelet l into new particle set.

This approach tends to favor lanelets with small estimated probability.
During resampling, most new particles are assigned to the prediction group their ancestor belonged to. A small percentage of particles can randomly mutate to the other group – or one of the others, in case of more groups. Like this, particles from a vehicle model or yaw rate source that currently fits the actual trajectory only badly are automatically reduced by being assigned lower weights in the measurement update and then being discarded in the resampling step. Whenever both models describe the vehicle's behavior equally well, the number of members of both groups will level out over time.

3.4.8 Lane Hypothesis and Availability

The output of the algorithm is the hypothesis on the currently used lane and the estimated probability mass function over all available lanes. To find the lanelet probability estimates, first all lanelet sum weights $\hat{p}_l$ are calculated according to Equation (3.69) and then the lanelet with maximum sum weight

$$l_{\text{hyp}} = \arg\max_l \hat{p}_l \tag{3.70}$$

is determined. In the evaluation, longitudinal positioning is not in the main focus and the particles may span over some short succeeding lanelets even with converged filter. In these cases, the probability estimate on the lanelet hypothesis $\hat{p}_{l_{\text{hyp}}}$ would not properly describe the lateral positioning. Therefore, the sum weights of lanelets directly before and after the hypothesis are added to the evaluation probability $\hat{p}_{\text{e},l}$. In the example particle distribution in Figure 3.16, the hypothesis is lanelet 4 and the corresponding evaluation probability estimate adds the probability estimates for lanelets 3 and 5:

$$\hat{p}_{e,4} = \hat{p}_4 + \hat{p}_3 + \hat{p}_5. \tag{3.71}$$

The same is performed for all neighbors of the hypothesis lanelet, i.e.

$$\hat{p}_{e,1} = \hat{p}_1 + \hat{p}_0 + \hat{p}_2 \tag{3.72}$$

for the left neighbors and

$$\hat{p}_{e,7} = \hat{p}_7 + \hat{p}_6 + \hat{p}_8 \tag{3.73}$$

for the right neighbors, and then continued in the same manner on further occupied lanelets starting with the remaining lanelets with highest sum weight.

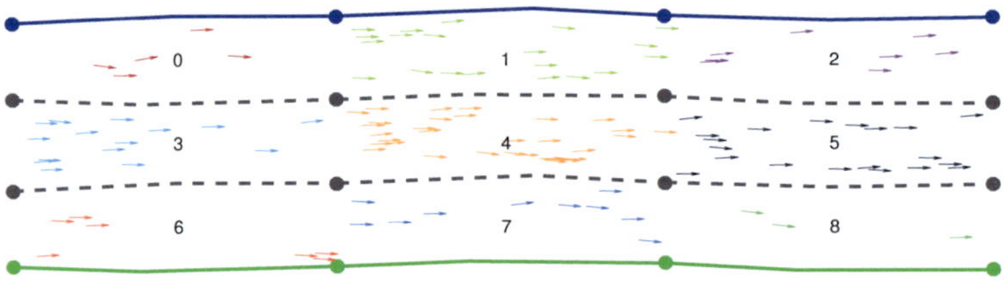

Figure 3.16: Example particle distribution on nine lanelets. Outer lanelet boundaries are shown in blue and green, respectively, and inner ones in dashed gray.

The value of the highest evaluation probability $\max_l \hat{p}_{e,l}$ can be seen as an indicator whether the algorithm has converged onto a single lane and whether its result is reliable. If this value is above a threshold probability p_{th}, the system is considered available. This threshold is of course an important design parameter as it allows for a tradeoff between high availability and low error probability. If the value is chosen very low, the system will be available in most cases but it is likely to accept a wrong solution in ambiguous situations. If the value is chosen too high, the error probability will drop but the system will rarely be available. Apart from the evaluation probability, also contradictions by observations of the blind-spot monitoring and front radar (Sections 3.4.6.1 and 3.4.6.2) can reduce the confidence in the result and set the system unavailable.

For the use in a lane-level navigation system, a probability mass function over possible lane indices in the given situation is required. The lanelet l with the highest evaluation probability $\arg\max_l \hat{p}_{e,l}$ defines the situation and the evaluation probabilities for all neighboring lanelcts are determined. Lane indices are counted from 0 on the left to $n_{\text{lanes}} - 1$ on the right. In the example of Figure 3.16, we assign probabilities to the left (0), middle (1) and right (2) lane. Aggregated particle weight on lanelets that are no neighbors of the lanelet hypothesis is not considered, but the probabilities are not re-normalized. Therefore the sum of the lane index-based probability mass function can be lower than 1.

3.5 Localization on Commercial Map

Current commercial maps used within automotive navigation systems are not designed for lane-level localization. While they feature the number of lanes and how they are connected over complex intersections and highway interchanges, geometry data is only available on road link level. This lack of lane-level geometry data is detrimental for a localization solution that relies on vehicle odometry. To deal with this, the localization scheme developed in Section 3.4 is adapted in several aspects as described in the following section.
First, map data is preprocessed automatically before using it for localization to exploit available hints on lane-level geometry as shown in Section 3.5.1. The filter itself is adapted in some general aspects that are described in Section 3.5.2. These include the use of additional data available in the map as well as the handling of unclear situations. Furthermore, besides a classical PF method described in Section 3.5.3, an alternative that combines an EKF for in-lane positioning and a PF for lane assignment is developed (Section 3.5.4).

3.5.1 Preprocessing Map Data

In the preprocessing step, the available map data is exploited to achieve an estimate for the lane-geometry and provide hints to the algorithm in unclear situations. The preprocessing consists of several individual steps described in the following.

3.5.1.1 Lane Offset to Centerline

According to digitalization guidelines, the road centerline lies in the middle of the available through lanes. Special lanes such as turn, acceleration, and deceleration lanes are not considered when determining the middle of the road. This is visualized in Figure 3.17: If turn lanes are equally distributed on the left and right side, the centerline is in the middle of the roadbed. If different numbers of turn lanes occur on both sides, the centerline does not describe the center of the roadbed. This offset of the centerline to the left is determined in a the first processing step as

$$o_i = \frac{n_{\text{special lanes left}} - n_{\text{special lanes right}}}{2}, \tag{3.74}$$

where left and right are the respective directions seen in digitalization direction.

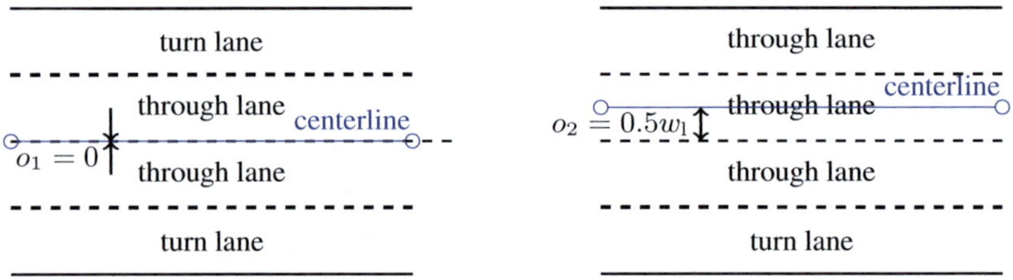

(a) Centerline in middle of roadbed. **(b)** Centerline with offset.

Figure 3.17: Examples for position of centerline in roadbed depending on type of lanes.

3.5.1.2 Shape Adaptation at Merges and Bifurcations

Link centerlines are digitized such that their respective end points share the same coordinate. This leads to a distortion of lane geometries as depicted for a simple case in Figure 3.18a: Two links, A with two lanes and B with one lane, merge into link C with three lanes. A corrected merge in this situation might look like the one shown in Figure 3.18b. Here, all three lanes from links A and B turn smoothly into link C. This can be achieved by moving the last shape point of the two merging links.

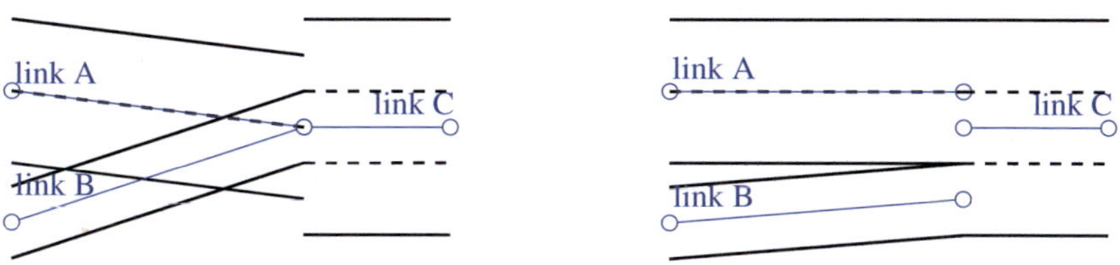

(a) Uncorrected. (b) Corrected.

Figure 3.18: Digitalization at merge of two links A, B into one link C.

However, in general the numbers of lanes do not match perfectly and some of the links may have a previously determined lane offset. Sometimes, bifurcations or merges happen at intersections and further links share the same node, or one link splits into three links at trifurcations. Hence, in a first step, possible links that make up a bifurcation or merge are collected by comparing the direction of the link with the most lanes to the directions of the other links. If the difference in lanes to and from the considered node is at most 1, the shape points of the merging or splitting links (A and B in Figure 3.18) are moved laterally to the direction of their first or last segment, respectively. Defining the direction corrected lane offsets

$$o'_i = \begin{cases} o_i, & \text{regarded node is reference node} \\ -o_i, & \text{else} \end{cases} \tag{3.75}$$

and using the numbers of lanes $n_{\text{lanes},i}$ of the three links $i \in \{A, B, C\}$, the displacement can be determined as

$$d_{\text{A}} = w_l \left(\frac{1}{2} \left(n_{\text{lanes,A}} - n_{\text{lanes,C}} \right) + o'_{\text{A}} + o'_{\text{C}} \right) \tag{3.76}$$

and

$$d_{\text{B}} = w_l \left(\frac{1}{2} \left(n_{\text{lanes,C}} - n_{\text{lanes,B}} \right) + o'_{\text{B}} + o'_{\text{C}} \right). \tag{3.77}$$

A possible third link at a trifurcation or merge of three links is not moved.

3.5.1.3 Spline Geometry

Commercial maps may be even more angular than lanelet maps. Therefore, OGH splines as described in Section 3.4.5.1 are used as well when evaluating distances to lane-markings. The map data provide entry and exit angles for many links to their neighbors that describe the orientation of a vehicle when entering of leaving a link. Of these angles, the one closest to the direction of the last segment is used for the direction of the tangent at the end of the spline. If no entry or exit angles are available, the direction of the first or last segment, respectively, is used.

3.5.1.4 Additional Lanes

Current infotainment maps contain turn lanes in situations such as highway interchanges and complex urban intersections. However, these data are not available everywhere. For many links, only the through lanes are attributed. In these cases, one cannot be completely sure whether a turn lane exists. To cope with this problem, *possible turn lanes* are introduced. Given that no information about turn lanes is available in the map for a certain link, it is checked into which direction one can turn from that link for all allowed travel directions. If it is possible to turn right, a possible turn right lane is added. For turn left maneuvers, a possible turn lane is only added for one-way links: The area of the possible turn lane is already occupied by the lanes for the oncoming traffic and the evidence for a turn lane is not strong enough to change the position of available through lanes.

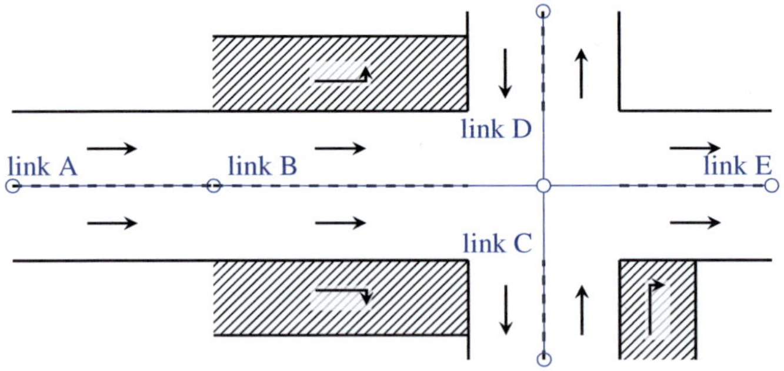

Figure 3.19: Intersection with possible turn lanes depicted as striped areas.

The concept is visualized in Figure 3.19: Without turn lane information for the five depicted links, one can assume that it is possible to turn left or right from link B. Link B is a one-way road and hence can have turn lanes on either side. From link C, one may only go straight or turn right. Accordingly, a turn lane might exist on the right side. Link D allows to turn left into E or go straight. However, it is a two-way road and no turn lane can be produced in this situation. For link A, it is only possible to move straight onto link B; in this case, no possible turn lanes are added. Section 3.5.2.3 describes how the localization filter deals with this uncertain type of lanes.

3.5.1.5 Lane-Width

For the map preprocessing, a constant lane-width is assumed for all road links. However, experience has shown that highway ramps and roundabouts feature wider lanes. Hence, they are treated as being 50% wider than the rest of the lanes. Furthermore, lanes of different roads are not the same width in practice. For example, highway lanes are wider than lanes in residential areas. For the combined PF/EKF localization, it is straightforward to estimate the current lane-width as described in Section 3.5.4.1. The pure PF scheme relies on a fixed lane-width estimate for all links.

3.5.1.6 Intersection Polygons

Lane geometry on intersections is not available in current infotainment maps. Instead, complex intersections are digitized with up to four nodes as depicted in Figure 3.20. The geometry that results from this digitalization and intersection internal links always being mapped as having only one lane, does not represent actual trajectories over the intersection well.

Using lane-marking or object observations together with this map in a localization filter, can lead to irritating results: For example, the ego-vehicle is located at the cyan triangle in Figure 3.20 and another vehicle is turning left over this intersection and is noticed by the radar sensor – depicted as the red point. Projecting the relative position of the observed vehicle onto the map will result in it being assumed somewhere in the middle of the intersection and off any mapped road for most particles close to the real position.

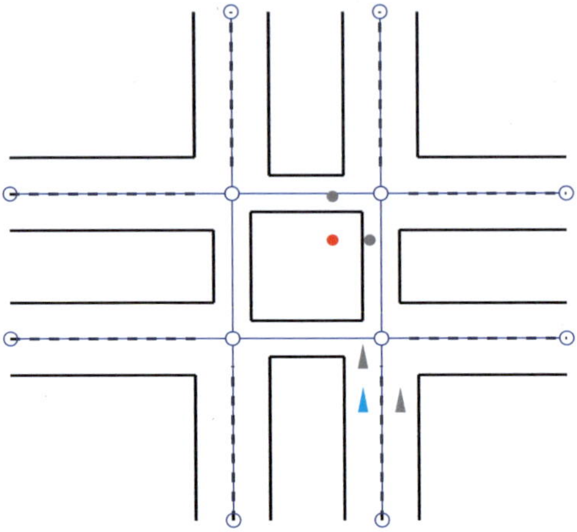

Figure 3.20: Centerlines at complex intersection and resulting lane geometry. Actual ego-vehicle position depicted as cyan triangle, an observed vehicle as red point. Two alternative hypotheses for ego-vehicle position depicted as gray triangles with corresponding projections of observed vehicle as gray points.

In turn, particles at the positions indicated as gray triangles might be favored as the observed vehicle is projected onto actual links. Hence, the wrong particles would be given a higher importance weight.

Similar to the actual geometry of lanes over intersections, the geometry of turn lanes around traffic islands ("bypasses") is often not represented well in current maps. They are digitized separately only if one side is longer than e.g. 25 m (HERE 2015) or 40 m (TomTom 2013). Again, both the ego-vehicle and observed other vehicles might travel over unmapped bypasses and distort the localization result.

As a solution, *intersection polygons* are introduced. They describe an area of an intersection in which hypotheses for the current position can travel freely following vehicle odometry, without being affected by measurement updates. To cover both the unknown lane geometry within an intersection and the possible existence of bypasses around the intersection, their borders are drawn into the oncoming links, compare Figure 3.21.

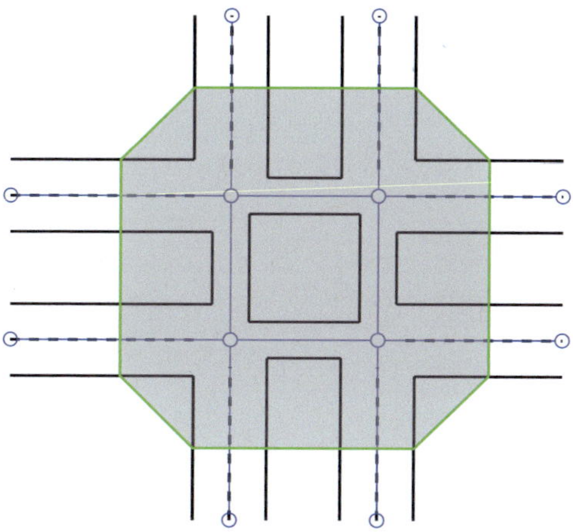

Figure 3.21: Complex intersection with intersection polygon.

3.5.2 Adaptations to Localization

Besides preprocessing the map, it is also necessary to adapt the localization algorithm to the different type of map and its accuracy. Information collected during preprocessing, such as possible turn lanes and intersection polygons need to be incorporated in a meaningful way into the filter. These adaptations are described in this section.

3.5.2.1 Link Transition

The map preprocessing steps toward better lane geometry described in Sections 3.5.1.1 and 3.5.1.2 already improve the continuity of mapped lanes. However, there remain situations where the correct transitions from one lane to another are unclear: In a transition from two to three lanes (Figure 3.22a) with both links having equal lane offset, the right lane of link A can lead to either the right or center lane of link B. Similarly, when the lane offset changes by half a lane but the number of lanes does not change (Figure 3.22b), the right lane of link A can lead to either lane of link B. For the left lane of link A, however, it is quite clear that it will connect to the left lane of link B.

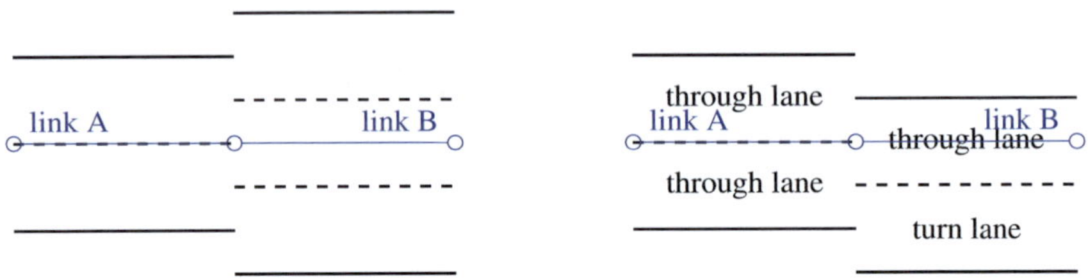

(a) From two to three lanes.

(b) A lane becomes turn lane.

Figure 3.22: Transitions between two links with unclear lane transitions.

This is handled based on the difference $|n_{\text{lanes},i} - n_{\text{lanes},j}|$ of the number of lanes and the offset difference $|o_i - o_j|$ between a source link i and a target link j. If the number of lanes is equal, $|n_{\text{lanes},i} - n_{\text{lanes},j}| = 0$ and the difference in offset $|o_i - o_j| < 1$, particles are moved laterally by the difference in offset $d = (o_i - o_j)\, w_l$. For bigger differences in offset, it assumed that turn lanes end or emerge and the geometry is acceptable. Hence, particles are not moved. For a change in number of lanes by $|n_{\text{lanes},i} - n_{\text{lanes},j}| = 1$ and a difference in offset of $|o_i - o_j| = 1$, particles are moved laterally by half a lane: $d = 0.5\,(o_i - o_j)\, w_l$. If the offset of both links is the same, $o_i = o_j$, particles are moved laterally randomly by half a lane, with 50% chance each for being moved to the left or right. The same happens if one link has two lanes more than the other, $|n_{\text{lanes},i} - n_{\text{lanes},j}| = 2$, and the offset differs by half a lane, $|o_i - o_j| = 0.5$. In all other cases, the particles are not moved.

3.5.2.2 Direction Measurement

Similar to the lanelet instance of the particle filter localization, the particles heading against the travel direction of their current link can be penalized. It can be legitimate to drive against the travel direction of the current link, e.g. when overtaking or going round another vehicle. However, this assumption is only reasonable if the two direction lanes are not physically divided. Hence, the penalty for going against intended travel direction is chosen differently, depending on whether a physical divider is present, i.e. depending on whether the link is one-way or not.

3.5.2.3 Additional Lanes

In Section 3.5.1.4, *possible turn lanes* have been introduced. The filter has to adapt to not certainly knowing whether these lanes actually exist. Hence, they are seen as drivable for both the ego-vehicle and other vehicles. At the same time, they are also considered as off-road and it is acceptable to observe stationary objects that are projected onto these additional lanes.

3.5.2.4 Stationary Radar Objects

In general, roads with two carriageways separated by a physical divider are digitized as two separate links. However, at many intersections, two-way roads split up to incorporate small traffic islands to accomodate traffic signs or traffic lights, compare Figure 3.23a. These traffic islands are often too small or not relevant enough to navigation to digitize both carriageways separately (Figure 3.23b).

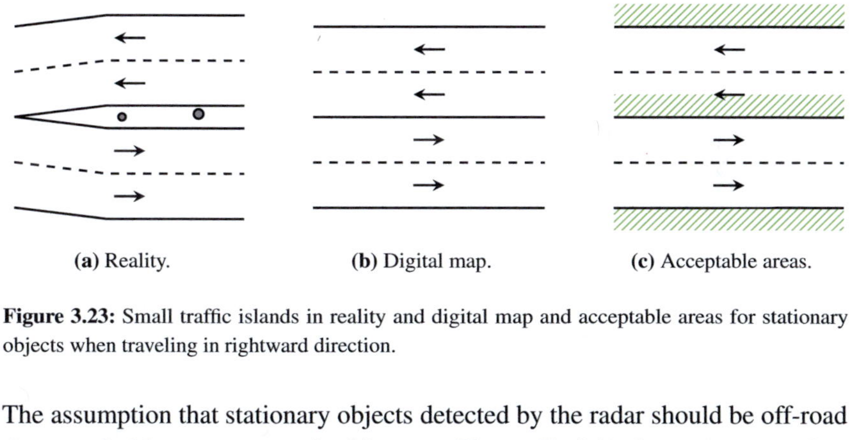

(a) Reality. (b) Digital map. (c) Acceptable areas.

Figure 3.23: Small traffic islands in reality and digital map and acceptable areas for stationary objects when traveling in rightward direction.

The assumption that stationary objects detected by the radar should be off-road does not hold true anymore in this case: The traffic island area is mapped as road. To handle this situation, not only stationary objects projected into off-road areas are acceptable, but also a patch on the leftmost lane of oncoming traffic, as depicted in Figure 3.23c for the case of moving in rightward direction.

3.5.2.5 Lane Connectivity Measurement Update

Current maps contain information on lane connectivity at complex intersections. This information describes which lane of a link before the intersection is connected to which lanes of which links after the intersection. Partially, intermediate links are included or this information is even available at lane configuration changes.
Small uncertainties or errors in longitudinal position estimation may lead to incorrect lane estimates after a turn over an intersection. This effect can be reduced by incorporating a measurement update step using lane connectivities. To achieve this, the current lane and link on which a particle is moving is stored. As soon as a particle reaches a new link – possibly over an intersection polygon –, it is evaluated whether the previous and new lane are connected. If not, the particle is assigned a reduced weight, i.e.

$$w_{\text{conn}}^{[i]} = \begin{cases} 1, & \text{lanes connected} \\ w_{\text{conn,min}}, & \text{else.} \end{cases} \tag{3.78}$$

3.5.2.6 Lane-Marking Type Measurement

The commercial map contains lane-marking types for some of the lane segments. These data can be used as a hint on the currently used lane: If the detected lane-marking type matches the one in the map, a lane is more likely than one without a match. However, the map and the lane-marking detection do not differentiate between the same set of lane-marking types. Also, experience has shown that some types are more easily confused than others. This leads to the weighting matrix for the two lane-marking types shown in Table 3.1. The most significant confusion is between solid / continuous markings and any broken / dashed type. Confusion between different broken / dashed types is significantly more likely and hence has a reduced impact on the particles' importance weight. In cases where the lane-marking types are not included in the digital map, a simplified scheme is used. It is based on the highway assumption that the outermost lane-markings are solid and inner ones are dashed. With the possible turn lanes introduced in Section 3.5.1.4, all types of lane-markings are accepted – i.e. the weight factor is 1 – as it is not sure whether these lanes actually exist.

Table 3.1: Weighting matrix for lane-marking types when data is available in map

Observed type		broken	contin-uous	broken narrow	block marking
Mapped type	dashed	1.0	0.9	1.0	0.99
	solid	0.95	1.0	0.95	0.95
	short dashed	1.0	0.9	1.0	0.99
	shaded blocks	0.99	0.9	0.99	1.0
	other types	1.0	1.0	1.0	1.0

Furthermore, a central line dividing a roadbed into two groups of lanes with different driving directions may be either solid or dashed. The resulting scheme is shown in Table 3.2.

Table 3.2: Weighting matrix for lane-marking types when no data is available in map for different positions relative to road boundary (r.b.) or road center

Observed type	broken	contin-uous	broken narrow	block marking
≥ 1 lane to r.b.	1.0	0.975	1.0	1.0
no lane to r.b.	0.99	1.0	0.99	0.99
possible lane to r.b.	1.0	1.0	1.0	1.0
line is center line	1.0	1.0	1.0	1.0

When applying the lane-marking weight it must be noted that the camera sees the road in front of the vehicle which might already be a different link in the map. Furthermore, the longitudinal accuracy of lane-marking types is low in the used map: type changes may be more than 20 m shifted from where they occur in reality. Hence, not only the lane-markings of the current link of each particle is considered but also previous and following links if they have the same number of lanes and the particle is close to the end of its current link.

3.5.2.7 Lane Hypothesis

Just like for the particle-filter based localization on lanelet maps, the output of this localization method is a hypothesis on the current lane and an estimated probability mass function over all lanes of the current link. However, the evaluation method described in Section 3.4.8 cannot be directly used for this filter: To avoid fluctuation of the lane hypothesis probability whenever transitioning from one link to another, the lane probabilities of the previous and following links are added to the lane probabilities of the current link.
For any lanelet, the succeeding and preceding lanelets can be easily determined and neighboring lanelets in general don't share any precursors or successors. In a commercial map with unclear lane geometry, the correspondences to lanes of other links are not certain, as depicted in Figure 3.22. Hence, a similar heuristic approach to the link transitions described in Section 3.5.2.1 is required.
If the succeeding or preceding lanelets are unclear, half of the lane probability of each possible successor or precursor, respectively, is added to the lane probability at the current link, comparable to the transition scheme in Section 3.5.2.1. Like this, a reasonable estimate for the lane probability mass function can be achieved and fluctuation at link transitions can be reduced.
Additionally to the probabilities for lanes, also a probability estimate for intersections is determined by summing the weight of all particles that are currently within an intersection polygon, i.e.

$$\hat{p}_{\text{intersection}} = \sum_{\boldsymbol{x}_i \text{ on intersection}} w^{[i]}. \tag{3.79}$$

3.5.3 Particle Filter Localization

For the PF approach, each particle stores its 2D-pose (Equation (3.6)). However, additional freedom of movement compared to the lanelet-based localization is required: While for the lanelet map and the used data set, it was a valid assumption that the vehicle is only driving within the boundaries of mapped roads. When using a commercial infotainment map, some lanes or bypasses around traffic islands might not be included in the map data. In Sections 3.5.1.4 and 3.5.1.6, possible turn lanes and intersection areas have been introduced. These need to be integrated into localization.

However, it is still possible that existing turn lanes are not in the map or bypasses are further away from the intersection point as expected. In these cases, it is necessary to allow particles to travel off-road. While the prediction step remains the same, the measurement update has to be changed: Off-road, there is no meaningful way to incorporate lane-markings as it is unclear to which lane boundary observed lane-markings may correspond. Similarly, it is questionable whether an off-road pose hypothesis is better when an observed vehicle is on-road. Hence, particles are not immediately discarded when leaving a mapped roadbed but instead are assigned minimal weight at every observation. This avoids off-road particles being favored over on-road particles but reduces their influence compared to particles on the road with every update.

The desired behavior at intersections is more clear: A particle shall move freely when on an intersection and neither be preferred nor penalized over a particle traveling on a normal road. Otherwise, the localization result might be distorted: When a group of particles with a certain extent along the road approaches an intersection, at some point some particles will be within the intersection polygon and others will still be outside. If now a weight is applied to the particles that favors either of the groups, the algorithm will converge to the favored part of the particle distribution – which is not necessarily the more correct part.

Therefore, an *intersection mode* is introduced that is enabled as soon as at least 10% of the particles are located within an intersection polygon. In intersection mode, no measurements updates that require knowledge of lane geometry or relationship between lanes are applied on any particle. This includes observations from BSM and updates using the digital map as well as the CWUS step based on lane-markings. Updates from radar objects are applied. Vehicles may be moving on any trajectory over the intersection, but also stationary structures like signposts or traffic lights can be positioned on parts of the intersection. Hence it is acceptable for both stationary and moving objects to be inside an intersection polygon.

3.5.4 Combination of Particle Filter and Extended Kalman Filter

Experience has shown that the geometric accuracy of current commercial infotainment maps leads to difficulties in tuning the described particle filter-based approach for lane-level localization on this kind of map. During localization on an accurate lanelet map, the filter relies on the accuracy of the vehicle orientation estimate as it allows tracking lane-changes even without detected lane-markings (Section 3.4). Even when interpolating the road geometry with a spline, the orientation of mapped road in a commercial map is not reliable and all observations regarding the orientation of the vehicle relative to the road need to consider this insecurity of at least a few degrees. Hence, the particle cloud is only kept together by the combined weight update and sampling step and quickly diverges when no lane-markings are detected or the vehicle enters an intersection. To fully cover the whole pose space, a much higher number of particles would be required.

However, as the CWUS step introduced in Section 3.4.5 already borrows from the Kalman filter for in-lane localization, it seems reasonable to divide the filter into two components for in-lane localization and lane-assignment that interact with each other as shown in Figure 3.24.

The digital map is preprocessed as before to describe intersection areas and heuristically reconstruct lane geometries. Based on observed lane-markings and the current vehicle speed, an EKF determines the lane-width and the vehicle's position and orientation within the lane (Section 3.5.4.1). This information, the preprocessed digital map, and the remaining environment sensor data are used to feed two particle filters: One performs the actual lane-assignment (Section 3.5.4.2) by restricting particles to mapped road links and another that handles intersection traversals (Section 3.5.4.3). Finally, the current lane-hypothesis is determined (Section 3.5.4.4).

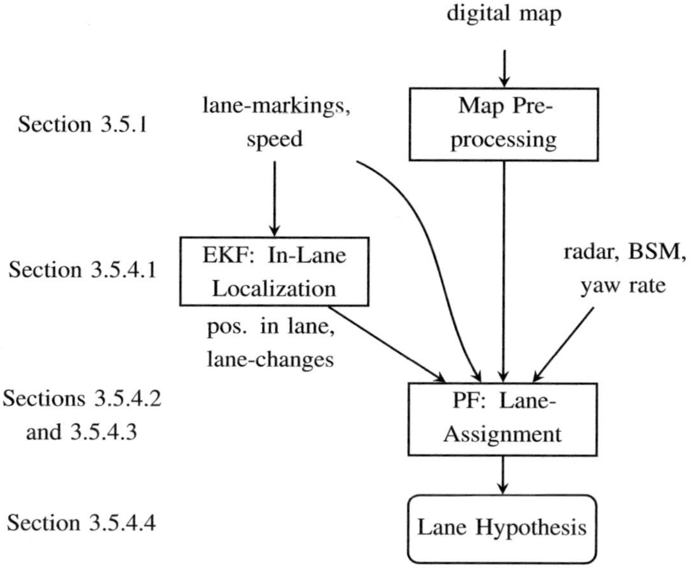

Figure 3.24: Overview of used combined particle-/Kalman filter algorithm and corresponding parts in this section.

3.5.4.1 In-Lane Localization

The in-lane localization component is an EKF that estimates the mean $\boldsymbol{\mu}_k$ and covariance matrix $\boldsymbol{P}_k$ of the system state

$$\boldsymbol{x}_{\mathrm{EKF},k} = (w_k, o_k, \alpha_k) \tag{3.80}$$

describing the lane-width w_k, offset to lane-center o_k, and orientation to lane-center α_k at time instance k. For the filter, the road is assumed to be straight and of constant width, leading to the prediction step

$$\bar{\boldsymbol{\mu}}_k = \boldsymbol{f}\left(\boldsymbol{\mu}_{k-1}, v_k, \Delta t_k\right) = \begin{cases} w_k & = w_{k-1} \\ o_k & = o_{k-1} + \sin\left(\alpha_{k-1}\right) \Delta t_k v_k \\ \alpha_k & = \alpha_{k-1}. \end{cases} \tag{3.81}$$

Linearization of this model leads to the Jacobian

$$\boldsymbol{J}_{F,k} = \left.\frac{\partial \boldsymbol{f}(\boldsymbol{x})}{\partial \boldsymbol{x}}\right|_{\boldsymbol{x}=\hat{\boldsymbol{x}}_{k-1}} = \begin{bmatrix} 1 & 0 & 0 \\ 0 & 1 & \cos(\alpha_k)\,\Delta t_k v_k \\ 0 & 0 & 1 \end{bmatrix} \tag{3.82}$$

that is then employed in the prediction of the covariance matrix (THRUN et al. 2005)

$$\bar{\boldsymbol{P}}_k = \boldsymbol{J}_{F,k}\boldsymbol{P}_{k-1}\boldsymbol{J}_{F,k}^{\mathsf{T}} + \boldsymbol{R}_k \tag{3.83}$$

with the process noise covariance matrix $\boldsymbol{R}_k$.
In the measurement update step, either the left or right lane-marking may be observed, leading to the two linear measurement models

$$\boldsymbol{z}_{\mathrm{L}} = \begin{bmatrix} d_{\mathrm{L}} \\ \alpha_{\mathrm{L}} \end{bmatrix} = \boldsymbol{J}_{H,k,\mathrm{L},z}\boldsymbol{x}_k + \boldsymbol{q}_{k,\mathrm{L}} = \begin{bmatrix} z & 1 & 0 \\ 0 & 0 & 1 \end{bmatrix} \begin{bmatrix} w_k \\ o_k \\ \alpha_k \end{bmatrix} + \boldsymbol{q}_{k,\mathrm{L}} \tag{3.84}$$

and

$$\boldsymbol{z}_{\mathrm{R}} = \begin{bmatrix} d_{\mathrm{R}} \\ \alpha_{\mathrm{R}} \end{bmatrix} = \boldsymbol{J}_{H,k,\mathrm{R},z}\boldsymbol{x}_k + \boldsymbol{q}_{k,\mathrm{R}} = \begin{bmatrix} z & -1 & 0 \\ 0 & 0 & 1 \end{bmatrix} \begin{bmatrix} w_k \\ o_k \\ \alpha_k \end{bmatrix} + \boldsymbol{q}_{k,\mathrm{R}}, \tag{3.85}$$

with the measurement noise $\boldsymbol{q}_{k,S}$ for the respective side $S \in \{\mathrm{L}, \mathrm{R}\}$ and $z \in \left\{-\frac{1}{2}, \frac{1}{2}, \frac{3}{2}\right\}$, depending on which lane-marking is observed: in different situations, the observed left lane-marking may be either the closest left ($z = \frac{1}{2}$) or right ($z = -\frac{1}{2}$) lane-marking or the second lane-marking on the left ($z = \frac{3}{2}$). The same occurs accordingly for the observed right lane-marking. Of course, z is in general not constant in time instance k and detection side S but these indices are omitted for readability. Hence, the observation error

$$e_{k,S}(z) = \|\boldsymbol{z}_S - \boldsymbol{J}_{H,k,S,z}\bar{\boldsymbol{\mu}}_k\| \tag{3.86}$$

is evaluated for the three discrete values of z and z is chosen to minimize this observation error

$$z = \arg\min_{z}\left(e_{k,S}(z)\right) = \arg\min_{z}\left(\|\boldsymbol{z}_S - \boldsymbol{J}_{H,k,S,z}\bar{\boldsymbol{\mu}}_k\|\right). \tag{3.87}$$

With the Kalman gain

$$\boldsymbol{K}_k = \bar{\boldsymbol{P}}_k \boldsymbol{J}_{H,k,S,z}^{\mathsf{T}} \left(\boldsymbol{J}_{H,k,S,z} \bar{\boldsymbol{P}}_k \boldsymbol{J}_{H,k,S,z}^{\mathsf{T}} + \boldsymbol{Q}_k \right)^{-1}, \tag{3.88}$$

the new state estimate μ_k and covariance estimate $\boldsymbol{P}_k$ can be found as (THRUN et al. 2005)

$$\boldsymbol{\mu}_k = \bar{\boldsymbol{\mu}}_k + \boldsymbol{K}_k \left(\boldsymbol{z}_S - \boldsymbol{J}_{H,k,S,z} \bar{\boldsymbol{\mu}}_k \right) \tag{3.89}$$

and

$$\boldsymbol{P}_k = \left(\boldsymbol{I} - \boldsymbol{K}_k \boldsymbol{J}_{H,k,S,z} \right) \bar{\boldsymbol{P}}_k \tag{3.90}$$

with $S \in \{\mathrm{L}, \mathrm{R}\}$.
Whenever $|o_k| > \frac{w_k}{2}$, the estimated in-lane position has changed to an adjacent lane. In this case, a lane-change signal is emitted that includes the information whether the lane has changed to the left or to the right. Furthermore, the state dimension o_k is reset to

$$o_k' = o_k - \operatorname{sgn}(o_k)\, w_k. \tag{3.91}$$

3.5.4.2 Lane-Assignment

As previously, following multiple hypotheses for the current lane and vehicle position is done with a particle filter-based approach. Particles can be either on a road link or on an intersection. Samples on a mapped road link are treated as follows.

System State For the lane-assignment on road links, the system state

$$\boldsymbol{x}_{\mathrm{RL},k} = (l_k, i_k, s_k, y_k) \tag{3.92}$$

is estimated with the current road link l_k, the current lane i_k, link segment s_k, and the offset along this segment y_k. Whilst this state looks four-dimensional, it contains essentially only 1.5 degrees of freedom: it describes a position in a two-dimensional space which is restricted to discrete lanes on links. i_k describes a discrete position lateral to the link while (s_k, y_k) contain the position along the link.

Initialization For the initialization, all links in the vicinity of the current predicted GPS position estimate are determined and assigned a probability according to their distance and orientation to the position estimate. Particles are then initialized among these links following these probabilities. For this, the closest point on the link to the GPS position estimate is determined. Then, the particle is randomly assigned to a lane. Finally, the positional inaccuracy is taken into account by moving the particles along the link randomly, following the assumed variance of the GPS error.

Transformation from Intersection Mode If a particle is not initialized at the very first start of the filter but is rather transformed from a sample leaving an intersection, initialization differs in these aspects: The particles are not assigned to the lanes following a uniform distribution. Instead, its stored lane from before the intersection is considered. If lane connectivities are available from the map, higher probabilities are assigned to lanes connected to the previous lanes while other lanes are only considered with reduced probability. If no lane connectivities are available and the lane structure is similar to the one before the intersection – compare Section 3.5.2.7 –, the previous probabilities are used to a smaller extent with a larger noise component. Lane-changes that have been detected during intersection traversal are applied afterwards.

Prediction As the particle state does not contain a rotational component, only speed estimates are used in the prediction step. Speed estimates are used similar to Section 3.4.3. However, a noise component with larger variance is used to account for the decreased accuracy of the map data.

Measurement Update Radar objects and BSM events are used in the measurement update step as described in Section 3.4.6. However, GPS pose estimates are also used in the update step to perform a kind of gating, comparable to CHAUSSE et al. 2005; F. LI et al. 2017, but also including the GPS heading estimate. For this purpose, for each particle the link segment orientation at the particle's position is used. It is defined as the linearly interpolated segment orientation between the current segment and the previous or next segment for all particles further than half a segment away from the link end. For particles close to the link end, the segment orientation from the map is used. It has to be noted that using the wheel speeds for the particle prediction and in the DR processing of the GPS receiver might lead to data incest problems (HÉRY et al. 2017). However, by only using the GPS position for eliminating particles far away from the current GPS position estimate, this occurs to not pose a problem. The absolute deviation

$$e_{\theta,k}^{[i]} = \left|\theta_l^{[i]} - \theta_{\mathrm{GPS},k}\right| \tag{3.93}$$

between the particle heading $\theta_l^{[i]}$ and the GPS heading estimate $\theta_{\mathrm{GPS},k}$ is used to assign a weight following

$$w_{\mathrm{GPS},\theta}^{[i]} = \begin{cases} 1, & e_{\theta,k}^{[i]} \leq 5^\circ \\ \exp\left(-\frac{\left(e_{\theta,k}^{[i]} - 5^\circ\right)^2}{2\sigma_{\mathrm{radar}}^2}\right), & \text{else.} \end{cases} \tag{3.94}$$

Allowing maximum weight for all deviations below 5° and reducing the weight only for bigger deviations allows to only punish particles on wrong links or segments and not introduce unwanted effects on particles on inaccurately mapped links.

Resampling As in Section 3.4.7, a systematic resampling step is applied whenever the effective sample size falls below a fixed threshold. However, the niching technique allowing particles on less populated lanes to survive, is not used.

3.5.4.3 Intersection Mode

Any particle that enters an intersection polygon is transformed to intersection mode and treated as follows.

System State In the intersection mode, the 2D-pose

$$\boldsymbol{x}_k = (x_k, y_k, \theta_k) \tag{3.95}$$

is estimated. This allows for free movement on the intersection where the vehicle is not restricted to mapped lanes.

Initialization Intersection particles are initialized based on the states of the road link particle filter and the in-lane position EKF when entering an intersection. For this purpose, first the current global pose estimate $\boldsymbol{x}_k^{[i]}$ of each particle is determined. Then it is moved laterally and rotated according to the variances in these states estimated by the in-lane position EKF. This allows to incorporate the uncertainty about the position and orientation within the lane with the current position estimate on discrete lanes in a straightforward matter.

Prediction Prediction is performed based on speed and yaw rate measurements as in the original particle filter described in Section 3.4.3.

Measurement Update As no lane structure is available on intersections, abstract measurements from BSM and lane-marking types are not incorporated. Furthermore, the structure of paved and drivable areas and stationary infrastructure in between is unclear. Hence, radar measurements are also not used. GPS is used in the same way as for the road link mode of this filter.

3.5.4.4 Lane Hypothesis

A lane hypothesis can only be determined based on particles on road links. In the first step, preliminary lane weights $\hat{p}'_{l,k}, l \in \{0, \dots, N-1\}$ are then determined at time instance k as in the PF approach in Section 3.5.2.7. However, if the current in-lane position estimate is close to the lane boundary and its variance is non-zero, the system cannot be certain about on which lane the vehicle is currently located. The same holds for arbitrary positions in the lane for large variance values, i.e. when the estimated Gaussian posterior spans over multiple lanes. Hence, in a second step, the previously determined lane weights $\hat{p}'_{l,k}$ are distributed over the existing lanes using the respective parts of the posterior of the in-lane position as depicted in Figure 3.25: The overall lane probabilities $\hat{p}_{l,k}$ at time k are the sums of highlighted areas under all in-lane position densities restricted to one lane l,

$$\hat{p}_{m,k} = \sum_{l=0}^{N-1} \hat{p}'_{l,k} \int_{x=lw_k}^{(l+1)w_k} \frac{1}{\sqrt{2\pi\sigma_{o_k}^2}} \exp\left(\frac{(x - o_k - (l+0.5)w_k)^2}{2\sigma_{o_k}^2}\right) \mathrm{d}x. \tag{3.96}$$

In the shown example with preliminary lane probabilities $\hat{p}'_{0,k} = 0.1$, $\hat{p}'_{1,k} = 0.7$, and $\hat{p}'_{2,k} = 0.2$ and an in-lane position estimate 1 m left of the center of the lane with a certain standard deviation, this leads to overall lane probabilities of $\hat{p}_{0,k} \approx 0.25$, $\hat{p}_{1,k} \approx 0.57$, and $\hat{p}_{2,k} \approx 0.15$. It can be seen that all preliminary lane probabilities influence the overall lane probability of their left neighbor in this scenario. If the in-lane position is accurately known, e.g. through reliable and repeated lane-marking detection, the influence of preliminary lane probabilities onto neighbor lanes is reduced.

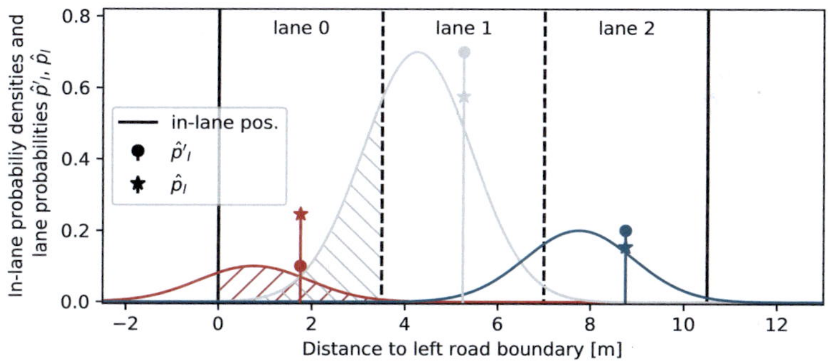

Figure 3.25: Influence of preliminary lane probabilities $\hat{p}'_{l,k}$ of lanes $l \in \{0, 1, 2\}$ onto the overall probability $\hat{p}_{0,k}$ of lane 0 for in-lane position estimate with non-zero variance. Probability densities depicted scaled by $\sqrt{2\pi\sigma^2}$ for better visibility.

4 Experimental Results

This chapter shows the performance of the presented localization solutions and puts them in comparison. The main aspects of the performance are the system availability and the error probability. In the first section, an introduction to the evaluation methods and metrics is given. Section 4.2 presents the performance results for the particle filter localization based on lanelet maps described in Section 3.4. Then, the results for the approaches based on commercial map data are given in Sections 4.3 and 4.4. Finally, a comparison between the presented algorithms puts them into context.

4.1 Evaluation Methods and Metrics

The main part of the performance evaluation within this chapter is based on offline evaluations of data logged from a test vehicle. Offline evaluations allow for parameter adjustment and repeated evaluation on the same data set. Especially with approaches that contain random aspects – such as the particle filter – it is required to determine whether the performance spread between different runs stays within reasonable bounds or is highly dependent on the realization of random variables. The used test tracks are described in Section 4.1.1. Section 4.1.2 gives an overview of the used data set for the evaluation. In Section 4.1.3, the evaluation logic is described: How is the ground truth determined and how is a correct lane assignment defined?

4.1.1 Test Tracks

Evaluation of the localization solution is restricted to the area covered by the used map. The particle filter-based localization described in Section 3.4 works on a lanelet map that was generated manually based on satellite images. This map covers two main tracks in Sindelfingen/Böblingen and Stuttgart and the connecting highways BAB 81, BAB 831, and B 14 and segments from L 1187 and L 1015. A total of around 2700 lanelets describe the area shown in overview in Figure 4.1.

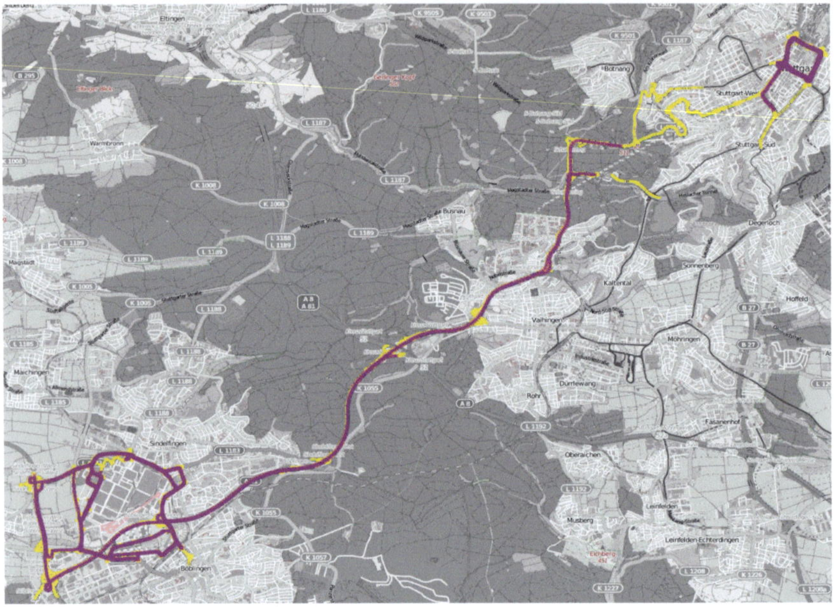

Figure 4.1: Lanelet map coverage overview, Sindelfingen in the bottom left corner, Stuttgart in the top right corner. The nodes that define the lanelet map geometry are depicted in yellow, some of the used track data overlayed in magenta. Background map © OpenStreetMap contributors.

The test track for suburban scenarios lies around the Mercedes-Benz plant Sindelfingen, see Figure 4.2. The urban scenarios are tested in downtown Stuttgart around the federal highways B 14, B 27, and B 27a, as shown in overview in Figure 4.3. This includes the major intersections at Arnulf-Klett-Platz, Gebhard-Müller-Platz, and Charlottenplatz that feature up to seven lanes per direction as well as the tunnel below Schlossplatz. While most of the map is based on satellite images, the tunnel segment was drawn using DR tracks.

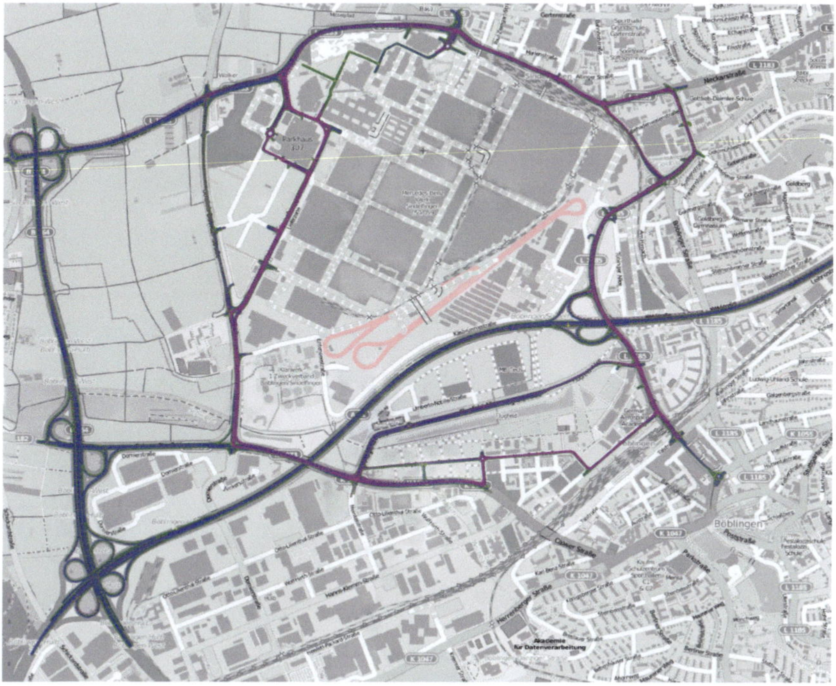

Figure 4.2: Lanelet map coverage Sindelfingen/Böblingen. Nodes of map in yellow, polyline segments (lanelet boundaries) in blue and green, some of the used track data in magenta. Background map © OpenStreetMap contributors.

4.1.2 Test Data

A total of around 6.5 hours or 230 km of logged data are used for the evaluation. The ratio of time spent on certain lane configurations, i.e. the number of lanes in driving direction, is shown in Table 4.1 for all data and for these three scenarios: These data contain around 80 minutes or 31 km on urban roads in downtown Stuttgart. Around 3.5 hours or 92 km have been driven on suburban roads in and around Sindelfingen and Böblingen. The rest of the data, some 107 km or 97 minutes, contain highways. Within these data, also some urban or suburban pieces are contained, for example tracks toward or from the highway, or intersections between exit and entry ramps when changing driving direction.

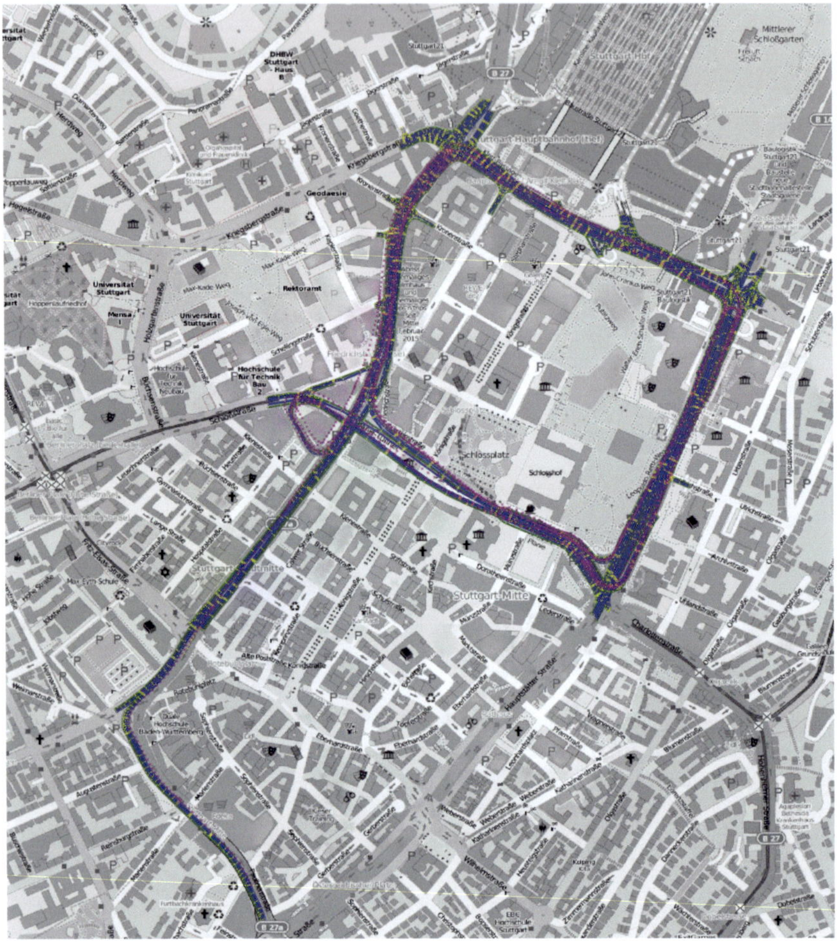

Figure 4.3: Lanelet map coverage Stuttgart. Nodes of map in yellow, polyline segments (lanelet boundaries) in blue and green, some of the used track data in magenta. Background map © OpenStreetMap contributors.

As the map contains mainly larger roads, i.e. urban sections of the federal highways B14, B27, and B27a, most data sets begin and the localization filter is initialized on multi-lane roads. This can increase ambiguity at initialization and increase convergence duration until the first certain localization result.

Table 4.1: Percentage of time spent on certain lane configurations

No. of lanes	1	2	3	4	5	6	7
% total	13.8	46.8	24.5	10.1	4.0	0.6	0.2
% urban	6.4	30.8	27.0	13.8	18.0	3.0	1.0
% suburban	18.8	51.3	20.8	9.0	0.1	–	–
% highway	6.0	28.9	38.9	24.6	1.7	–	–

Especially in urban scenarios, lane-markings are not always visible. Table 4.2 gives an overview of the amount of time that lane-markings are available in the used data set. In the highway scenario, both lane-markings can be detected in a vast majority of the cases (82%) and in less than 10% of the time, no lane-marking is observed. Less markings are available when driving in builtup areas. In 25% to around 28% of the time, no lane-markings are detected in the urban and suburban scenarios. During these unavailabilities, the algorithm has to perform localization based on data from other sensors.

Table 4.2: Percentage of time with available lane-markings

Available markings	only left	only right	both	neither
% total	11.4	11.7	55.0	21.9
% urban	14.8	14.7	42.0	28.4
% suburban	13.0	13.6	48.4	25.0
% highway	4.9	4.5	82.0	8.7

4.1.3 Ground Truth and Evaluation Logic

One aspect of the performance evaluation is the error probability, i.e. the percentage of time or distance in which the localization solution is available but the estimated lane is not correct. Due to the lack of a highly accurate DGPS system and a reliably geo-referenced map, the ground truth data required for this evaluation was hand-labeled based on video data, supported by GPS measurements. For localization on a lanelet map, the ground truth contains the correct lanelet with a precision of around 0.25 seconds. However, due to ambiguous situations at lanelet transitions or imperfect GPS positioning, the noted determined timestamp might not be as accurate. Hence, during performance evaluation, the lanelet hypothesis at a certain time instance t is determined as described in Section 3.4.8, and then all correct lanelets between $t - 0.5\,\mathrm{s}$ and $t + 0.5\,\mathrm{s}$ are looked up in the ground truth data. If the lanelet hypothesis is equal to any of the ground truth lanelets within that 1 second interval or any direct precursor or successor of those, it is considered correct. Performance evaluation can be performed based on time or distance traveled. While the qualitative result of both options is very similar and both can be used to differentiate between a good and bad algorithm, some quantitative differences can occur. The evaluations in this work are based on the time passed. If the lanelet hypothesis at t_k is correct, the complete time interval $t_k - t_{k-1}$ will be considered correct – and accordingly for false and unavailable lane assignments.

4.2 Particle Filter Localization on Lanelet Map

This section shows results of the particle filter-based localization described in Section 3.4 in a simulated scenario in Section 4.2.1 as well as on real-world data from suburban, urban, and highway scenarios in the following sections.

4.2.1 Simulation

While its performance on actual vehicle data and a digital map of the real world is the qualifying criterion for the quality of a localization algorithm, it is also interesting to examine its behavior in a simulation. Like this, sensor and map inaccuracies can be controlled and it is easier to reason certain design decisions and evaluate convergence behavior. This section shows how the particle filter approach performs in a simulation. It compares a traditional importance weight update step with the proposed combined weight update and sampling. Then, the influence of moving and stationary radar objects is evaluated before the localization is tested on an erroneous map.

4.2.1.1 Simulation Setup

The artificial signals in the simulation mimic the ones from vehicle log files in their interfaces and frequency. As the simulation shall mainly evaluate the influence of environment sensors on the localization, the velocity measurements describe a forward movement with constant velocity, the yaw rate measurements a straight trajectory. GPS data gives positions around the center lane of the map, moving from one end of the map forward with the same velocity as the odometry measurements; a circular uniform distribution with a radius of 15 m around the measured position is used for the simulations. The GPS heading is oriented in movement direction.
All scenarios include visual lane-marking detection that detects a constant distance of 2 m and angle of 0° to the left and right marking. The lane-marking type is not used. In Section 4.2.1.3, different data from the radar sensor are provided:

1. no radar objects
2. vehicles on ego- and first neighboring lanes
3. vehicles on all lanes

The used map contains three parallel straight lanes. In the scenarios used in Sections 4.2.1.3 and 4.2.1.4, a fourth lane and a digitalization error are added, respectively.

4.2.1.2 Traditional Weight Update vs. Combined Weight Update and Sampling

As motivated in Section 3.4.5, it is important that a filter does not converge towards one hypothesis if the sensor data does not contradict the other hypotheses. Hence, the scenario described in the toy example in Section 3.4.5.2 is evaluated in more detail: The simulated vehicle is traveling on a perfectly straight three-lane road and lane-marking detection provides a constant distance and angle to the left and right lane-marking. No sensor measurement contradicts any of the lanes and therefore the lane probabilities should stay around their initial values.

Performing this experiment with traditional importance weight update steps leads to the results shown in Figure 4.4. For $N_\mathrm{P} = 100$ particles, there are cases for each lane where its probability converges to 0 or 1. The median probability for the center and right lane are closer to 0 than to $\frac{1}{3}$. The used particle set is too small and the results are quite random.

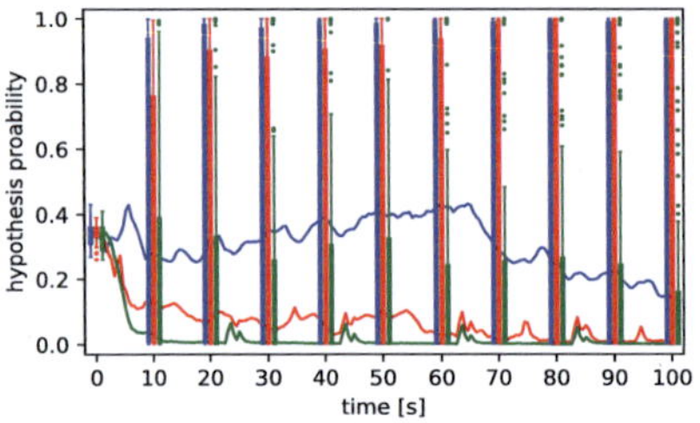

(a) 100 particles.

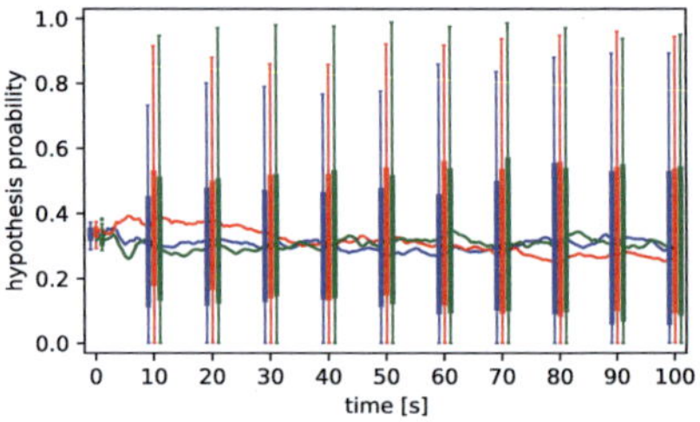

(b) 500 particles.

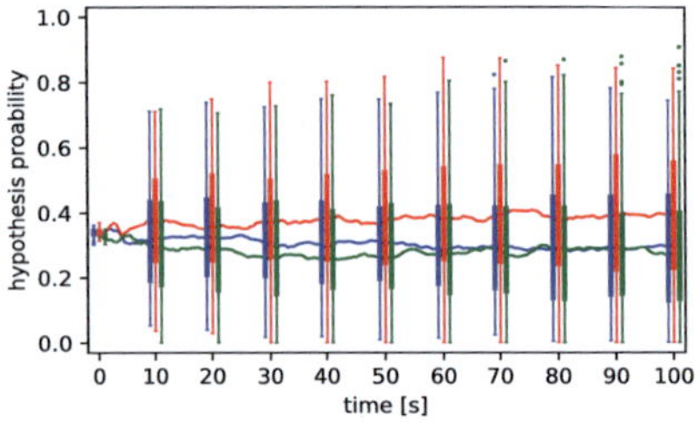

(c) 1000 particles.

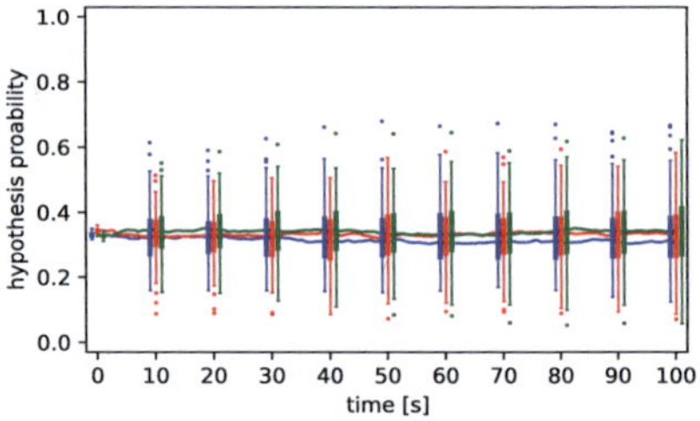

(d) 4000 particles.

Figure 4.4: Lane hypothesis probabilities in 100 repetitions over time with traditional weight update for different N_P. Left lane depicted in blue, center lane in red, right lane in green. Median probability as line. Boxplot whiskers with $1.5\times$ interquartile range. Figures (b) and (d) taken from RABE and STILLER 2017.

When increasing the number of particles, the variation between the repetitions decreases and the median probability gets closer to $\frac{1}{3}$. For 4000 particles, all lane probabilities stay between 0.25 and 0.4 for half of the repetitions but in few repetitions, the lane probabilities can still reach 0.1 and 0.7.

With a CWUS step and a Gaussian assumption for the intermediate belief and the likelihood, the median lane probability stays around $\frac{1}{3}$ for all used particle set sizes, see Figure 4.5. Already with 100 particles, a variation comparable to the one achieved with 4000 particles using a traditional weight update can be achieved. The reproducibility of the results is significantly higher and with 1000 particles, the lane probabilities stay within the $[0.25, 0.4]$ corridor for almost all repetitions.

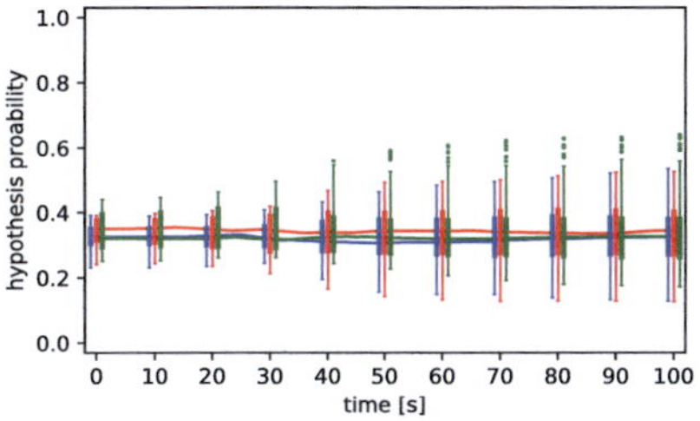

(a) 100 particles.

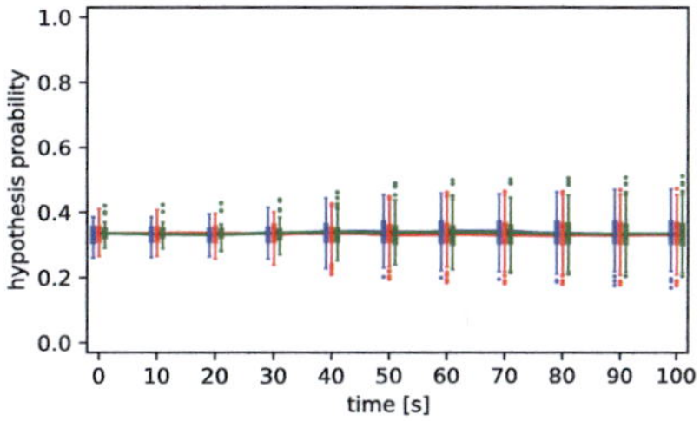

(b) 200 particles.

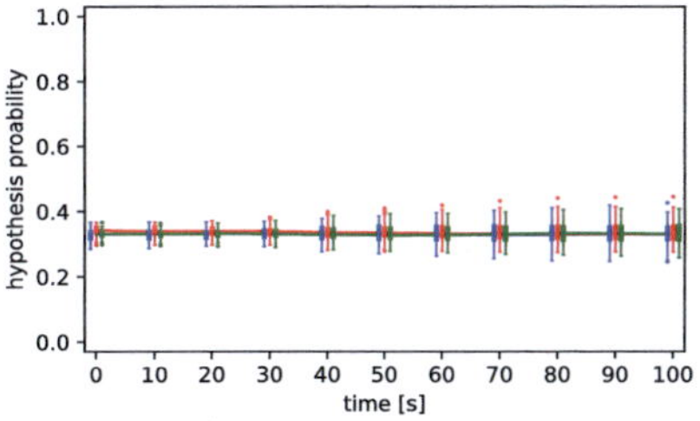

(c) 500 particles.

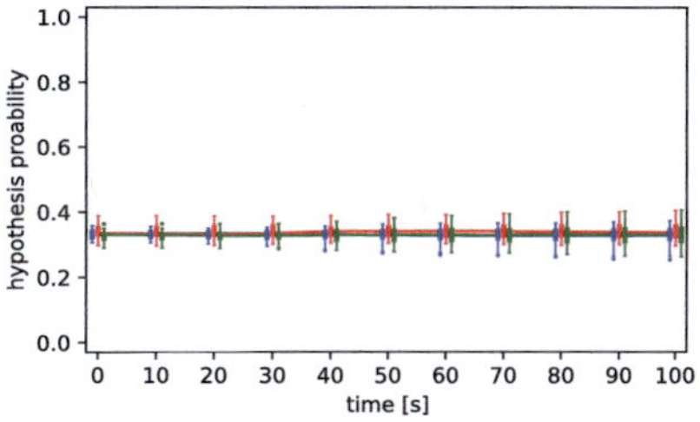

(d) 1000 particles.

Figure 4.5: Lane hypothesis probabilities in 100 repetitions over time with Gaussian CWUS. Figure (c) taken from RABE and STILLER 2017.

A similar result can be achieved under the assumption of mixture distributions for the intermediate belief and the likelihood, as shown in Figure 4.6.

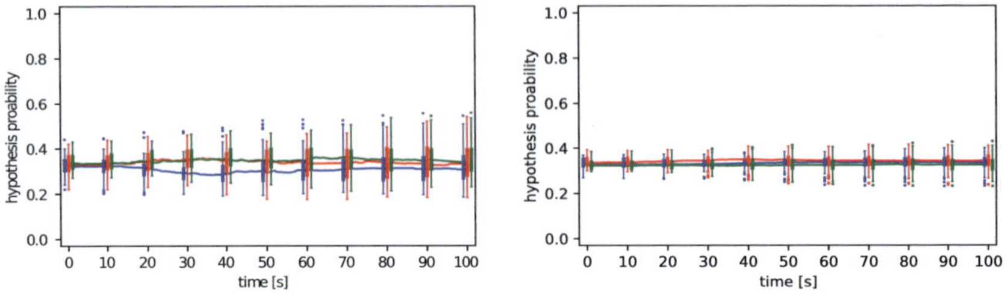

(a) 100 particles.

(b) 500 particles.

Figure 4.6: Lane hypothesis probabilities in 100 repetitions over time with mixture CWUS. Figure (b) taken from RABE and STILLER 2017.

4.2.1.3 Influence of Moving Radar Objects

Three-lane road In the next step, other vehicles detected by radar are also incorporated. The ego-vehicle is driving on the middle one of three lanes and moving objects on all three lanes are available. Hence, the probability of the middle lane is expected to converge towards 1, whereas the other two converge towards 0. Again, both proposed schemes for incorporating lane-markings are compared at different numbers of particles N_{P}.

The results are shown in Figure 4.7. Using a traditional weight update and $N_{\mathrm{P}} = 100$ particles, the median result converges as expected but in a noticeable amount of cases, the probability of the middle lane tends to zero. When increasing the number of particles to $N_{\mathrm{P}} = 500$, the results are much more reproducible and only in some cases, the probability of the middle lane is small after 10 and 20 seconds. After at least 30 seconds, the filter result is as expected.

When in turn using a CWUS step with either Gaussian or mixture assumption, the result is already reproducible with 100 particles and very similar to the result with 500 particles. The particle efficiency for a desired result reproducibility has been increased.

Four-lane road, vehicles on first neighbor lane With the same observations as in the previous scenario but on a four-lane road, the situation is ambiguous as the vehicle might be on either of the two center lanes.

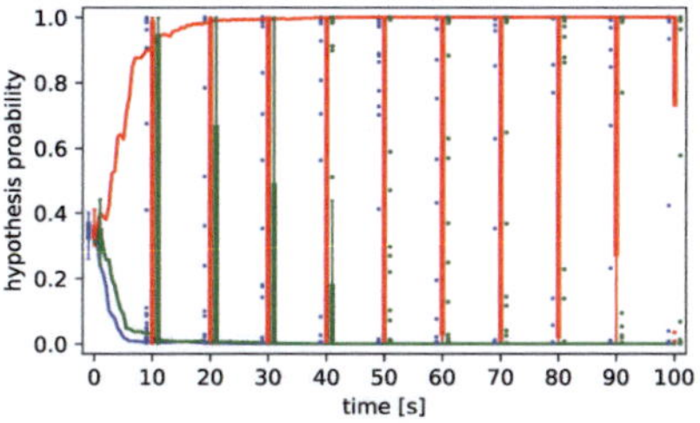

(a) Traditional update, $N_\mathrm{P} = 100$.

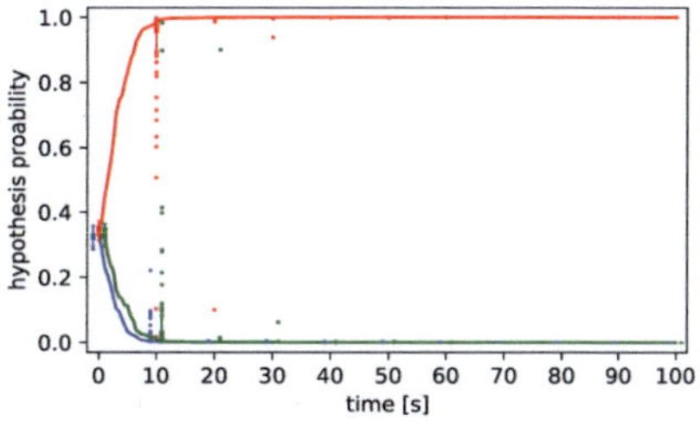

(b) Traditional update, $N_\mathrm{P} = 500$.

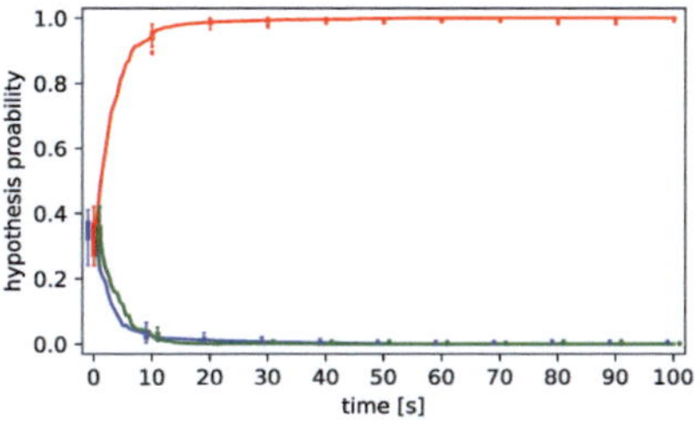

(c) Gaussian CWUS, $N_\mathrm{P} = 100$.

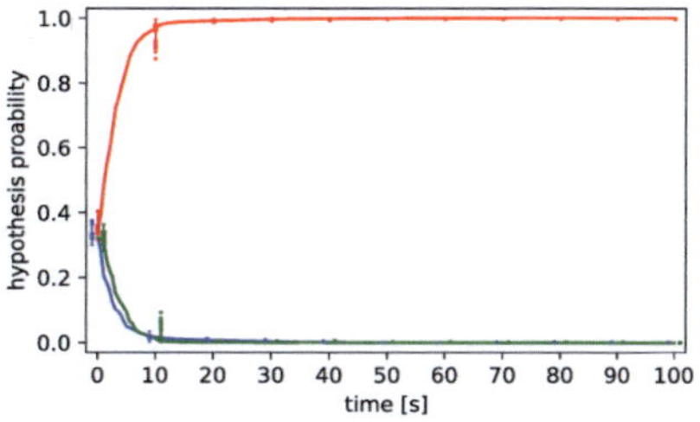

(d) Gaussian CWUS, $N_\mathrm{P} = 500$.

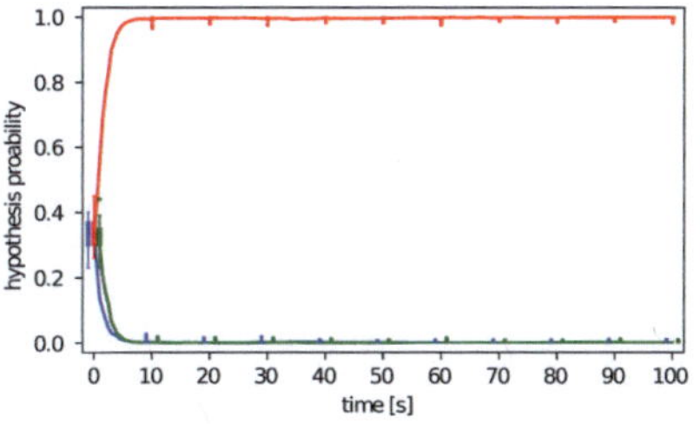

(e) Mixture CWUS, $N_\mathrm{P} = 100$.

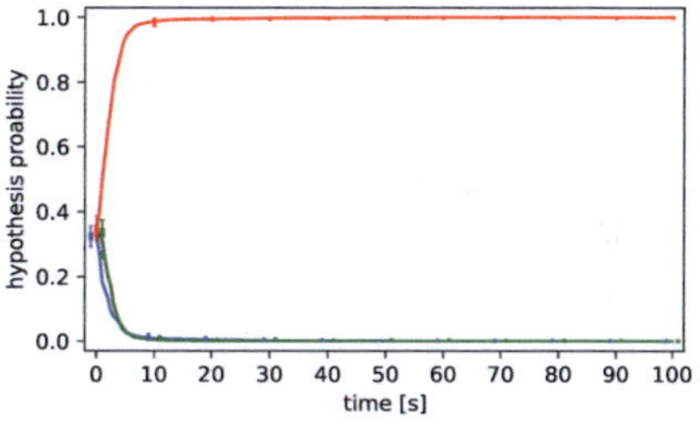

(f) Mixture CWUS, $N_\mathrm{P} = 500$.

Figure 4.7: Lane hypothesis probabilities in 100 repetitions with detected vehicles for $N_\mathrm{P} \in \{100, 500\}$ and different methods to use lane-markings. Road with three lanes.

In some vehicle architectures, only objects on the direct neighbor lanes can be transmitted. In this case, the expected result is that the probabilities for the outer lanes converge to zero and those of the inner lanes to 0.5.

The results in Figure 4.8 show that again, a traditional update in combination with a small particle set leads to irreproducible behavior: With $N_\mathrm{P} = 100$, the median probability of the inner left lane is the only one not converging to zero and for each of the four lanes, there exist situations where its probability converges to 1. This effect can be reduced with $N_\mathrm{P} = 500$, but the estimated probabilities for the inner lanes still vary between 0 and 1.

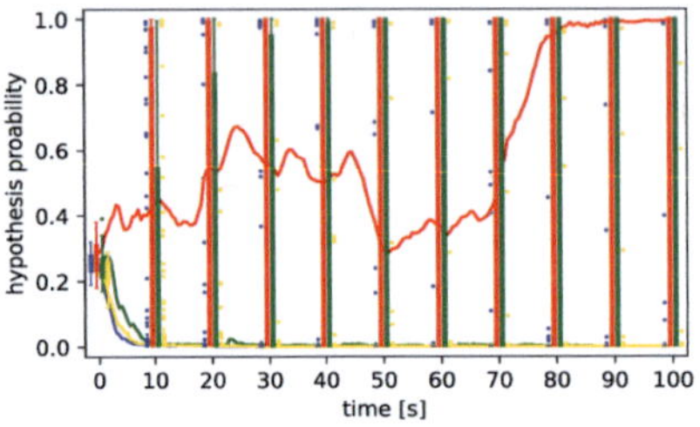

(a) Traditional update, $N_\mathrm{P} = 100$.

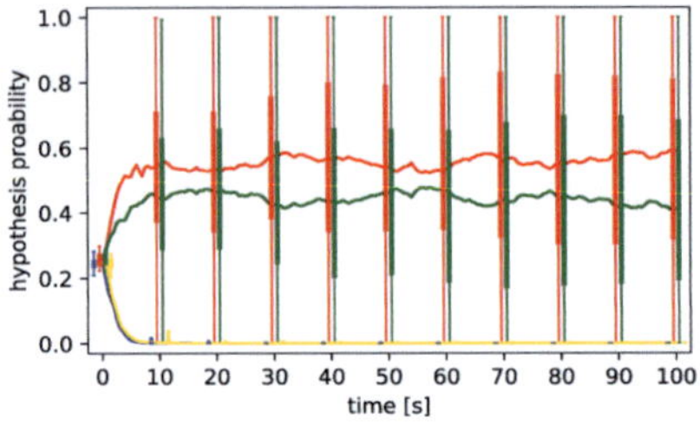

(b) Traditional update, $N_\mathrm{P} = 500$.

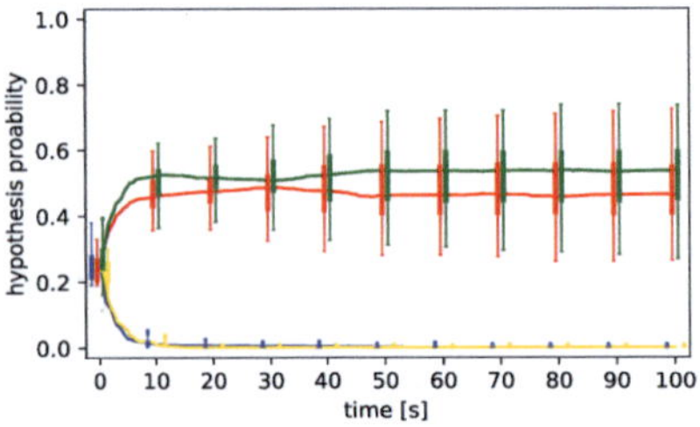

(c) Gaussian CWUS, $N_\mathrm{P} = 100$.

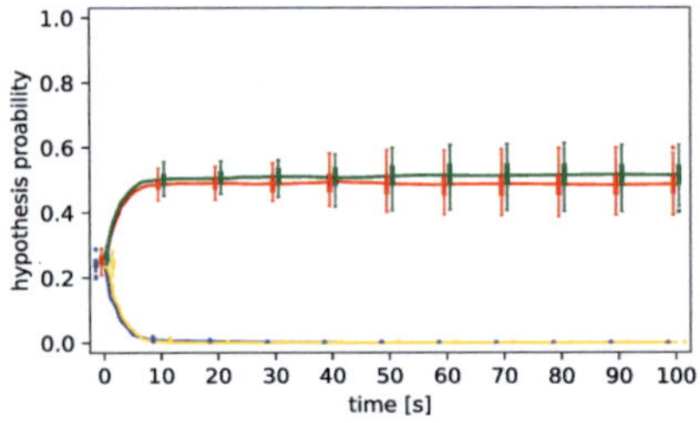

(d) Gaussian CWUS, $N_\mathrm{P} = 500$.

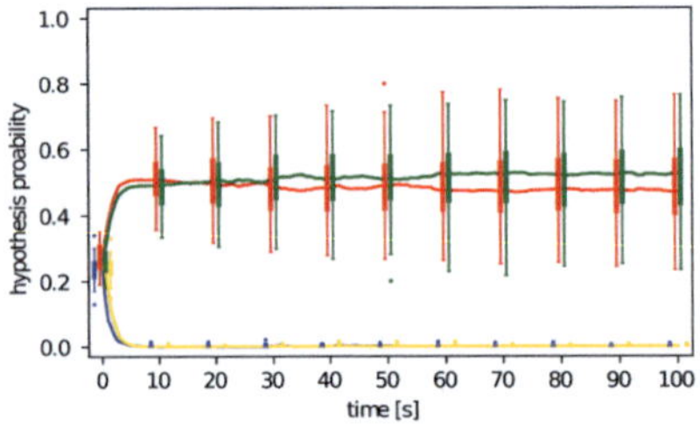

(e) Mixture CWUS, $N_\mathrm{P} = 100$.

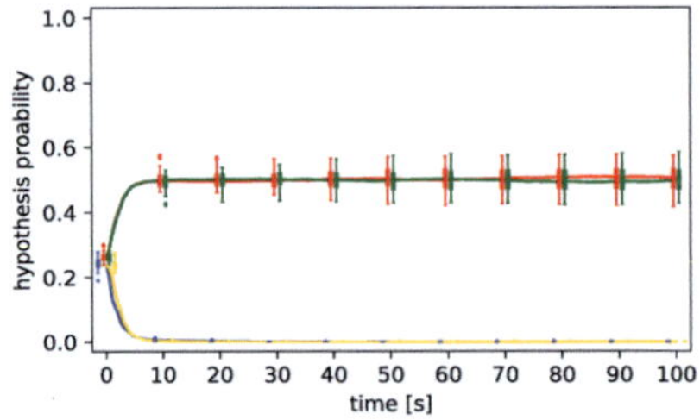

(f) Mixture CWUS, $N_\mathrm{P} = 500$.

Figure 4.8: Lane hypothesis probabilities in 100 repetitions over time with detected vehicles on the direct neighboring lanes for $N_\mathrm{P} \in \{100, 500\}$ and different methods to use lane-markings. Road with four lanes. Outer left lane depicted in blue, inner left lane in red, inner right lane in green, outer right lane in yellow.

Employing a CWUS step instead, the results are already more reproducible with $N_\mathrm{P} = 100$. For $N_\mathrm{P} = 500$, the probabilities of the inner lanes only rarely leave the corridor $[0.4, 0.6]$ within the 100 second simulation. This result agrees with the two previous ones.

Four-lane road, vehicles on all lanes If additionally vehicles on the second neighboring lane can be detected, one can expect the filter output to converge to the correct lane. In this case, the ego-vehicle is on the inner left lane and vehicles on the outer right lane are detected less frequently as on the direct neighboring lanes.
The results are depicted in Figure 4.9 and show that with a traditional weight update used for lane-marking detection, the particle set size $N_\mathrm{P} = 100$ is still too small to produce useful results. With $N_\mathrm{P} = 500$, the filter converges to the correct lane after around 60 seconds in almost all cases but the variation before is significant.
With the proposed CWUS method, $N_\mathrm{P} = 100$ is enough to provide reliable results after around 20 seconds. Increasing the sample size to $N_\mathrm{P} = 500$ allows for a faster convergence and more reproducibility already after 10 seconds. Again, the results are comparable for both Gaussian and mixture assumptions for intermediate belief and likelihood.
Results comparable to these can be achieved with a combination of BSM information – which only gives information about the direct neighbor lanes – and detected guardrails or road boundaries – which can reduce ambiguity on four- or more-lane roads.

4.2.1.4 Influence of Map Errors

Digital maps are in general not perfect. The digitalization rules of mapping companies require 5 m absolute position accuracy and 1 m relative position accuracy for maps for Advanced Driver Assistance System (ADAS) applications (HERE 2015; TomTom 2016). As the absolute position accuracy is smaller than the assumed GNSS error, no performance loss is expected at this point.

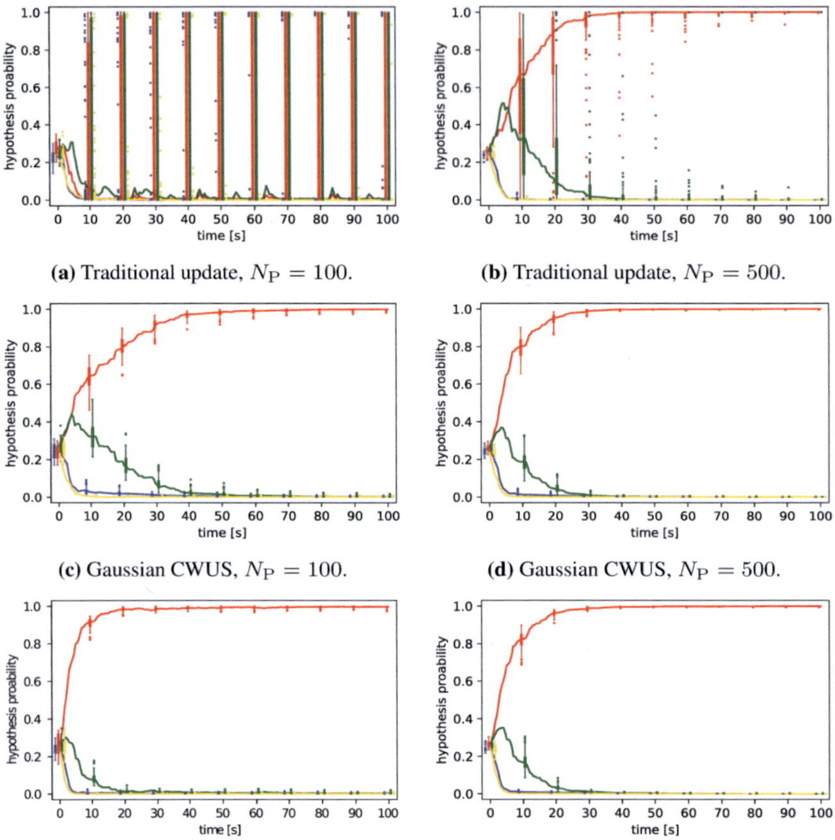

(a) Traditional update, $N_\mathrm{P} = 100$.

(b) Traditional update, $N_\mathrm{P} = 500$.

(c) Gaussian CWUS, $N_\mathrm{P} = 100$.

(d) Gaussian CWUS, $N_\mathrm{P} = 500$.

(e) Mixture CWUS, $N_\mathrm{P} = 100$.

(f) Mixture CWUS, $N_\mathrm{P} = 500$.

Figure 4.9: Lane hypothesis probabilities in 100 repetitions over time with detected vehicles on all lanes for $N_\mathrm{P} \in \{100, 500\}$ and different methods to use lane-markings. Road with four lanes.

However, the relative position accuracy of the used lanelet map is better than the quality requirements posed by mapping companies. Hence, it is interesting to find out whether relative positioning errors influence the localization solution. To evaluate this, the three-lane map used for the previous experiments has been changed such that the whole road exhibits a bend of a whole lane-width – around 4 m – to the right over a length of 14 m. The sensor data is the same as before, i.e. the vehicle is moving in a straight line and detecting left and right lane-markings, and, in the second case, other vehicles. The vehicle passes the mapped bend after around 54 seconds. Again, a traditional importance weight update step is compared to the proposed CWUS step. 100 repetitions have been carried out and $N_{\mathrm{P}} = 500$ particles have been used.

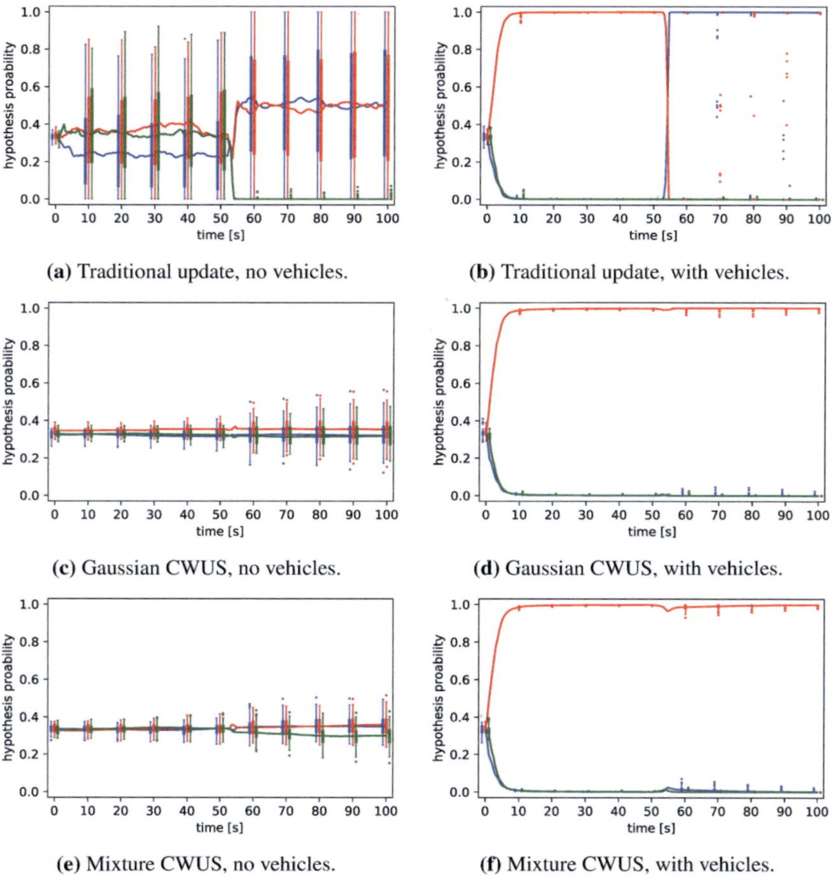

(a) Traditional update, no vehicles.

(b) Traditional update, with vehicles.

(c) Gaussian CWUS, no vehicles.

(d) Gaussian CWUS, with vehicles.

(e) Mixture CWUS, no vehicles.

(f) Mixture CWUS, with vehicles.

Figure 4.10: Lane hypothesis probabilities in 100 repetitions over time for $N_{\mathrm{P}} = 500$, with and without detected vehicles, and different methods to use lane-markings. Road with a bend. Figure (a), (c), and (e) taken from RABE and STILLER 2017.

Only lane-markings In the first case, only lane-markings are observed. Using a traditional update, as shown in Figure 4.10a, the lane hypothesis probabilities vary a lot between different repetitions before passing the bend. When passing the bend, in most cases all particles on the right lane (green line) are lost and the left and center hypotheses share the particle weight. However, the variation between repetitions increases further and in some cases, only one of the hypotheses survives.
When employing a CWUS step, the particles follow the mapped lane more closely and all lane hypotheses still exist after the bend. With both a Gaussian (Figure 4.10c) and a mixture CWUS step (Figure 4.10e), the variation between iterations increases slightly after the bend. The median result stays around $\frac{1}{3}$, even though the right lane's probability is slightly reduced when using a mixture likelihood.

Lane-markings and other vehicles In the second scenario, observations of other vehicles are used as well. The ego-vehicle is moving on the middle lane. With a traditional lane-marking update, as in Figure 4.10b, most particles on the middle lane move straight ahead at the bend and end up on the left lane when passing the bend. In few repetitions, some particles remain on the middle lane and keep the correct hypothesis alive. However, with the used process noise, the filter has no chance to adapt to the map error.
Both variants of the proposed CWUS step manage to keep most particle weight on the correct hypothesis despite the map error, compare Figures 4.10d and 4.10f. The median probability of the middle lane is slightly reduced at the bend and the variation of the left and middle lane probabilities slightly increased afterward. We can see that employing a CWUS step for lane-marking detection allows the filter to handle this kind of map error.

4.2.2 Optimal System Design

The achieved error rate and availability on real world sensor data and maps is the qualifying criterion for a localization solution. To achieve this, a suitable parameter set needs to be found: The used set of sensor data needs to be determined, the several inputs need to be weighted according to their reliability, and likelihood functions and process noises need to be modeled. Throughout development, this has been performed on a reduced data set by comparing results achieved with different parameters.

A parameter that can be treated separately is the threshold p_{th} above which a lane hypothesis is accepted and the system is considered available. Intuitively, it allows for a trade-off between error rate and availability: A low threshold increases the availability but also leads to a higher error rate. In turn, the error rate can be reduced with a high threshold – at the cost of lower availability.

To choose this parameter, one can either define a target error probability, e.g. 0.5% of time, or minimize a cost function that weights the error rate r_{e} stronger than the unavailability $1 - r_{\mathrm{a}}$, such as

$$C = 10r_{\mathrm{e}} + (1 - r_{\mathrm{a}})\,. \tag{4.1}$$

The aggregated result of the system using a Gaussian CWUS step is shown over the threshold parameter p_{th} in Figure 4.11. As expected, a higher availability comes at the cost of a higher error rate: When choosing $p_{\mathrm{th}} = 0.8$, an average error rate of $\bar{r}_{\mathrm{e}} = 0.29\%$ of time can be achieved at an average availability of $\bar{r}_{\mathrm{a}} = 93.8\%$. Reducing this threshold to $p_{\mathrm{th}} = 0.6$, the availability increases to $\bar{r}_{\mathrm{a}} = 97.5\%$ on average, while the error rate rises to $\bar{r}_{\mathrm{e}} = 0.57\%$.

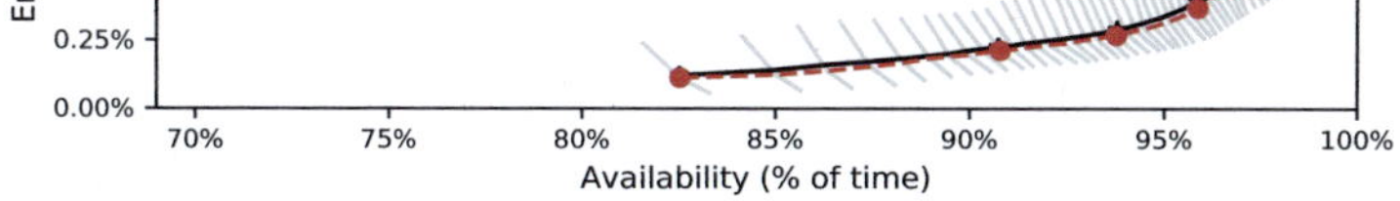

Figure 4.11: Error rate over availability of time for different threshold probabilities $p_{\mathrm{th}} \in \{0.5, 0.51, \ldots, 0.99\}$ and 100 repetitions with different random seeds. Mean and median with markers at $p_{\mathrm{th}} \in \{0.5, 0.6, \ldots, 0.9, 0.99\}$, percentiles over repetitions as whiskers from respective median.

If a lane-precise navigation function requires a mean error rate of at most $\bar{r}_{\mathrm{e}} = 0.5\%$, this can be achieved with threshold parameter $p_{\mathrm{th}} = 0.64$. In this case, the system is available $\bar{r}_{\mathrm{a}} = 96.8\%$ of the time. Minimizing the cost function in Equation (4.1) leads to the similar threshold

$$p_{\mathrm{th}} = \arg\min_{p_{\mathrm{th}}} \left(10 r_{\mathrm{e}} + (1 - r_{\mathrm{a}})\right) = 0.65, \tag{4.2}$$

where an error rate of $\bar{r}_{\mathrm{e}} = 0.45\%$ at an availability of $\bar{r}_{\mathrm{a}} = 96.5\%$ can be achieved. If not noted otherwise, the following results use the threshold parameter of $p_{\mathrm{th}} = 0.64$.

4.2.3 Behavior in Different Scenarios

The goal of the described approach is to reliably determine the currently used lane in many situations that are relevant for navigation, including urban downtown scenarios. To see whether any scenario poses special challenges for the algorithm, the result has been evaluated separately in real-world data collected in drives in urban, suburban, and highway scenarios described in Section 4.1.1.

Suburban Scenario The suburban data set based on drives in Sindelfingen and Böblingen is handled very well by the algorithm, compare Figure 4.12a. For the given threshold parameter $p_{\mathrm{th}} = 0.64$, a mean error rate $\bar{r}_{\mathrm{e}} = 0.4\%$ can be achieved with an availability of $\bar{r}_{\mathrm{a}} = 96.8\%$. Even in the 95th percentile, the error rate does not exceed $r_{95,\mathrm{e}} = 0.82\%$. With a higher threshold parameter $p_{\mathrm{th}} = 0.94$, the average error rate decreases to $\bar{r}_{\mathrm{e}} = 0.1\%$ with an availability of still $\bar{r}_{\mathrm{a}} = 90.1\%$.

Urban Scenario The higher number of lanes per direction and more complicated road structure makes the downtown scenario in general more difficult. However, the proposed system excels in this discipline and reaches an average error of $\bar{r}_{\mathrm{e}} = 0.37\%$ of time for the given threshold parameter – at an availability of $\bar{r}_{\mathrm{a}} = 96.2\%$, as shown in Figure 4.12b. The fifth worst error rate in 100 repetitions lies at $r_{95,\mathrm{e}} = 0.74\%$. Again, it is possible to reduce the average error rate to $\bar{r}_{\mathrm{e}} = 0.1\%$. In this scenario, this requires a threshold of having

$p_{\text{th}} = 94\%$ of the particle weight on one lane and yields an availability of $\bar{r}_{\text{a}} = 80.9\%$.

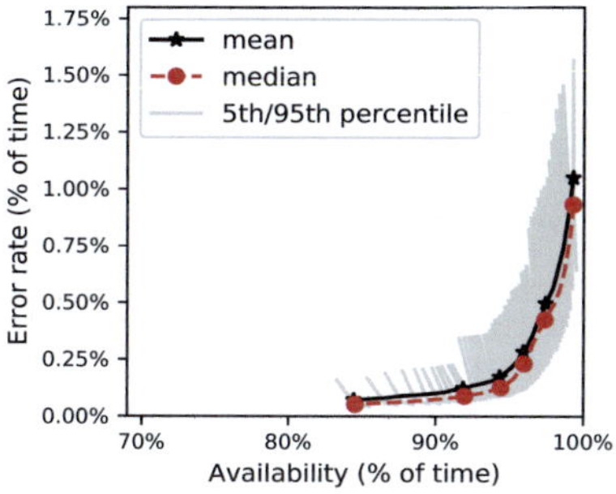

(a) Suburban scenario.

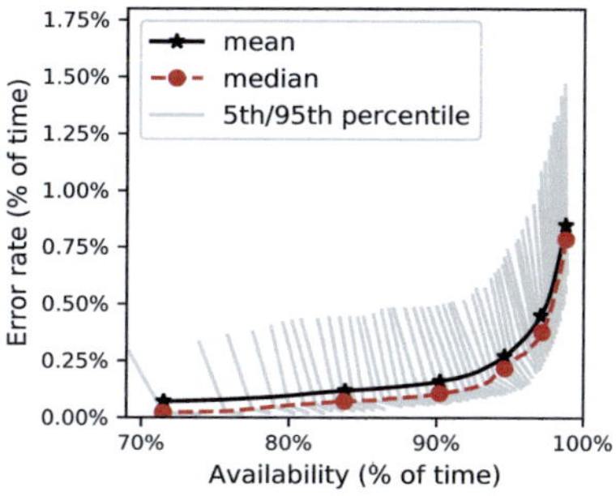

(b) Urban scenario.

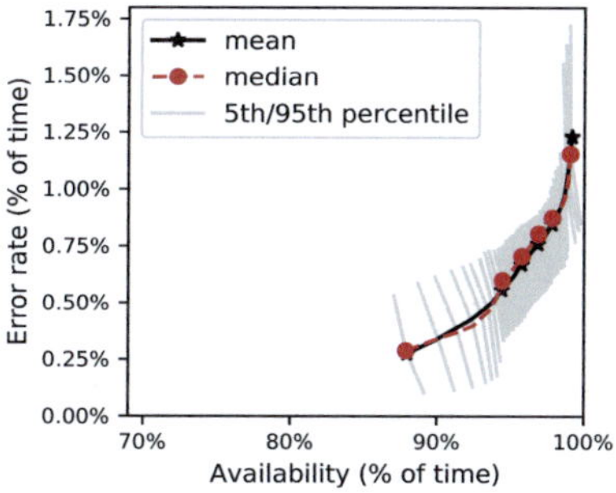

(c) Highway scenario.

Figure 4.12: Error rate over availability of time for different scenarios and different threshold probabilities p_{th} and 100 repetitions. Depiction as in Figure 4.11

Highway Scenario Lane-level localization has been tested on highways with a more straightforward histogram filter (MAYER et al. 2014). Therefore, the parameter set for this algorithm has been optimized for performance in urban situations. This leads to the performance on highway scenarios depicted in Figure 4.12c: With the used threshold parameter, the average error rate is $\bar{r}_{\text{e}} = 0.81\%$ at an availability of $\bar{r}_{\text{a}} = 97.5\%$. Yet, the results are reproducible with different random seeds and the 95th percentile of the error rate lies at $r_{95,\text{e}} = 0.91\%$.

In all three scenario types, lane estimates occur that are supported by around 50-60% of the particle weight but do not agree with the ground truth lane. In the highway scenario, however, even when increasing the lane probability required to accept a lane hypothesis up to 99%, the mean error rate does not decrease below a residual of $\bar{r}_\text{e} = 0.28\%$. This means that in some situations almost all particles are located on the wrong lanelet. Analysis of the remaining situations with wrong lane assignment shows two main error causes:

- *Longitudinal error*: The longitudinal position of the particle set is not directly corrected by any observation. GPS is only used for initialization and, during the localization, for estimating the yaw rate bias. Opposed to urban and suburban situations, this error can accumulate over longer distances and lead not only to a spread in position, but in some cases also to a position offset. This offset then transforms to a lateral offset after the first intersection after exiting the highway.

- *Erroneous sensor data*: In one situation in the data set, the railing of a highway ramp throws a shadow onto the road in front of the vehicle and is interpreted as a lane-marking for some hundred meters. The lane-marking observation is not robust to an error over such a long time and the particle set changes to a wrong lane. Error propagation effects lead to the lane-estimate only being corrected after a significant distance has been traveled and hence has noticeable influence on the overall performance in this scenario.

Influence of Number of Lanes Apart from the location of the current road and its speed limit, also the number of lanes in driving direction may have an influence on the localization performance. In Figure 4.13, the accuracy and availability are depicted for different lane configurations. As the ego-vehicle was located on one- to three-lane roads for around 85% of the used data, it is understandable that the overall result is very close to the the 0.4% to 0.5% error achieved on these segments. The increased error rate for four-lane roads is due to the detected railing that already lead to increased error rates for the highway scenario and is now responsible for almost half of the false lane assignments on four-lane roads in the evaluated data set. Within the evaluated data, the localization works successfully on roads with more than four lanes: Average error rates between 0.12% (five lanes) and 0.22% (seven lanes) can be achieved. It shall be noted, however, that the underlying sample for five- to seven-lane roads is rather small.

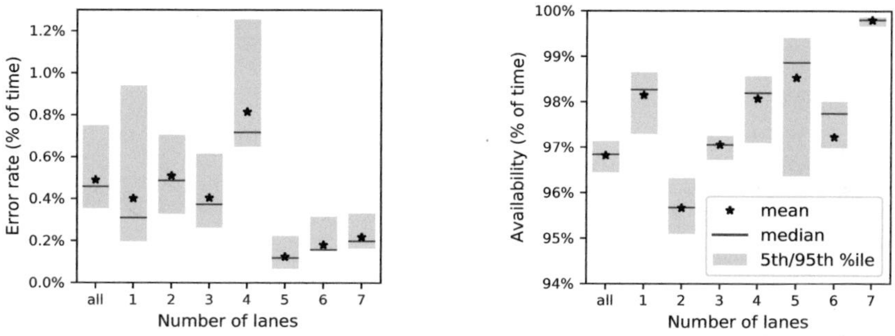

(a) Error rate. **(b)** Availability.

Figure 4.13: Error rate and availability over number of lanes for 100 repetitions with different random seeds.

The achieved availabilities vary slightly with means between $\bar{r}_\mathrm{a} = 95.7\%$ (two lanes) and $\bar{r}_\mathrm{a} = 98.5\%$ (five lanes). Only seven-lane roads yield even $\bar{r}_\mathrm{a} = 99.8\%$, mainly because the filter has already converged before the road splits up before an intersection in the few cases that the road becomes that wide.

4.2.4 Convergence Behavior

Apart from the overall availability, it is also relevant how long the system takes from startup to the first useful lane-assignment output. This duration is heavily dependent on the lane configuration at initialization and hence on the used data set. A histogram over 100 repetitions of the used data set is shown in Figure 4.14. In the suburban and highway scenarios, the algorithm converges to a trustworthy solution in less than a second in a couple of situations. For nine of the ten data sets in which this happens, the particle filter is initialized on one-lane roads, such as small suburban roads or highway ramps. These convergence durations correspond to distances driven of less than 7 meters, mostly between 1 and 2 meters.

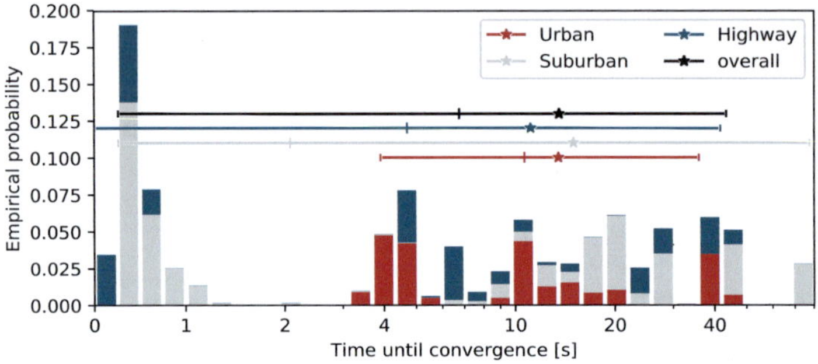

Figure 4.14: Time from initialization to first available localization for different scenarios (stacked) for 100 repetitions with different random seeds. Mean duration depicted as asterisk, median with 5th and 95th percentile as errorbars for each scenario. Linear scale below 1 second, logarithmic above.

The maximum convergence duration observed is 80 seconds in a suburban file. Here, the filter is initialized on a two-lane road shortly before a red traffic light where the car waits for around 70 seconds. On the segment before the traffic light and during the waiting, no observations leading to filter convergence can be observed.

In total, the average convergence duration is 13.5 seconds or 69 meters. In 95% of the cases, the filter delivers a result in less than 43.4 seconds or 199 meters. The average durations vary slightly for the different scenarios between 11.1 (highway) and 15.0 (suburb) seconds, which corresponds to 53 (suburb) to 94 (urban) meters. The difference is larger in the median convergence duration with around 2.1 seconds (suburb) and 10.7 seconds (urban) or 19 meters (suburb) and 66 meters (urban). It can be seen that the number of lanes at initialization and the traffic situation and observations after initialization have a significant influence on filter convergence.

When the convergence period is not considered in the overall performance, the availability rises to 98.5% for the chosen threshold parameter $p_{\mathrm{th}} = 0.64$. As less total time is considered, also the ratio of time with erroneous lane assignment rises slightly from 0.49% to 0.5%. The result is shown for different threshold probabilities in Figure 4.15. The initial unavailability before the first reliable localization result contributes more than half of the total unavailable time for many relevant values of p_{th}.

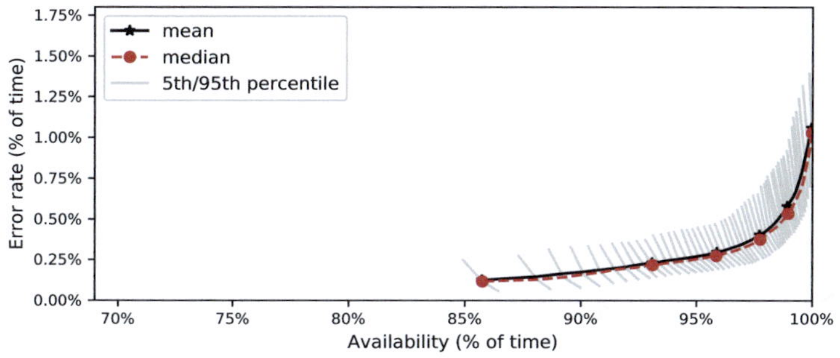

Figure 4.15: Error rate over availability of time for different threshold probabilities and 100 repetitions with different random seeds when not counting convergence to total time. Depiction as in Figure 4.11.

4.2.5 Comparison of Proposed Method

To rank the proposed aspects of the localization algorithm, their respective performance is compared with instances where they are disabled and a simpler alternative is used. In the following, the CWUS approach described in Section 3.4.5 is replaced by a traditional weight update step. Then, the influence of the mixture distribution assumptions on intermediate belief and likelihood (Section 3.4.5.3) is evaluated. Furthermore, the effects of the proposed preprocessing or filtering step on radar objects (Section 3.3) and the use of multiple particle sets with different prediction models (Section 3.4.3) are evaluated.

4.2.5.1 Combined Weight Update and Sampling

As seen in the simulations in Section 4.2.1.2, using a CWUS step for incorporating detected lane-markings increases the reproducibility of the result as well as the chance for the correct hypothesis to stay alive over differently seeded repetitions. A CWUS step should yield improved performance and lower variation within the results. Figure 4.16 shows the results of this comparison to a system with traditional weight update based on lane-markings. The error rate with CWUS (0.49%) is significantly lower than without (12.6%) at comparable availability values. Furthermore, the error rate variation between 90% of the repetitions is 0.4% with CWUS and 2.3% without.

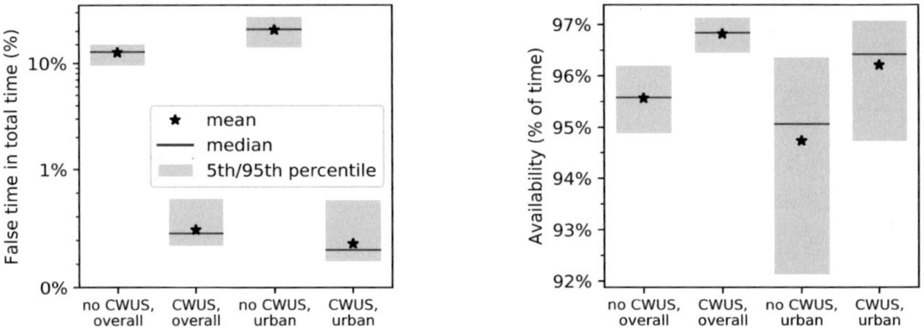

(a) Error rate.

(b) Availability.

Figure 4.16: Comparison between performance with and without combined weight update and sampling: Error rate and availability for 100 repetitions with different random seeds for complete data set and urban scenario.

The difference between both approaches increases even more when looking specifically at the downtown data set: Here the traditional weighting leads to 20.5% false lane assignments as opposed to 0.37% with the proposed CWUS step. In both cases, the variation in the availability is slightly increased without CWUS. This is because without CWUS, the correct hypothesis can only be favored if at least one particle happens to be close to the correct position within that respective lane.

4.2.5.2 Robust Combined Weight Update and Sampling

As shown in the previous section, employing a CWUS step for integrating detected lane-markings into the filter leads to a significant performance gain. However, it can be reasoned that the Gaussian assumptions on the intermediate belief and the likelihood do not model their actual shapes well. Hence, the approach has been extended in Section 3.4.5.3 to support mixtures of Gaussian and uniform distributions.

Through the iterative nature of the used EM algorithm for estimating the intermediate belief, the computational requirements have been increased. But does the increased complexity also improve localization performance? And does the seemingly more accurate model depict the actual particle distribution better than the plain Gaussian?

To find out, the results using a Gaussian assumption are compared to both approaches to mixture assumptions – one with determining the belonging of particles to either part of the mixture likelihood independently for each measurement, and one with an additional state whether the lane-marking measurements are reliable using a Markov chain.

A comparison of the results on the whole data set and in the urban scenario is shown in Figure 4.17. For this evaluation, an outlier probability of $\pi_{\mathrm{m}} = 0.25$ was used. One can notice an increased error rate of around 0.7% over all data when using the independent mixture approach, while the error rate using the Markov mixture approach (0.51%) does not differ significantly from the result with normal assumption (0.49%). In turn, the availability is slightly reduced for both mixture versions (96.5% and 96.4%, respectively) compared to the simple approach (96.8%). This is mainly due to the system trusting the lane-markings less – despite the lane-markings being a reliable data source within the used data set.

In the urban scenario, the Markov mixture approach shows a slightly reduced error rate (0.31% compared to 0.37%) but at the cost of an increased variation within the repetitions with different random seeds. This can be due to the lane-markings or the mapped lane boundaries being a little less reliable in downtown – but at the same time, the system does not always favor the particles with the correct lane-marking reliability state. Given the variation between the 100 repetitions, the difference in availability is insignificant.

In the suburban and highway data sets, the algorithm using the normal assumption for the intermediate belief outperforms the versions with the mixture assumptions – leading to the best performance on the complete data set. Hence, while the normal assumption is not a correct model for the distribution of the particles before the measurement update, it still leads to the best localization performance.

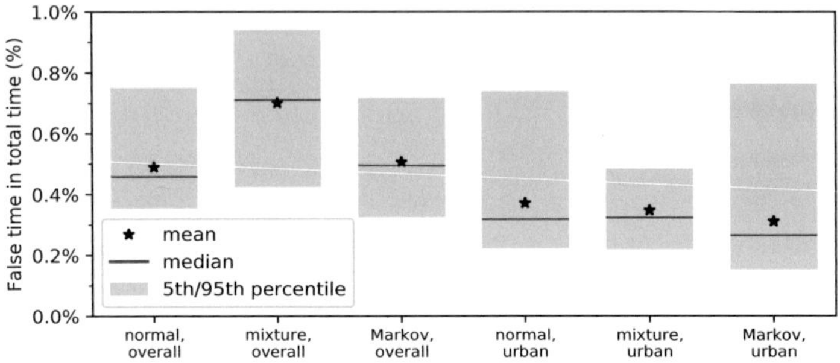

(a) Error rate.

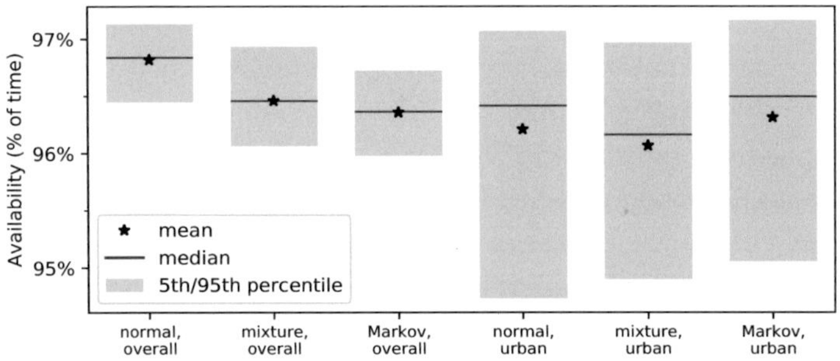

(b) Availability.

Figure 4.17: Comparison between performance for different combined weight update and sampling settings: Normal assumption and mixture assumption with independent and Markov modeling. Error rate and availability for 100 repetitions with different random seeds for complete data set and urban scenario.

However, if the data is corrupted, the robust CWUS step has its advantages (RABE and STILLER 2017). To emulate corrupt lane-markings, 10% of the measured distances to the lane-markings have been randomly selected at run-time and have been replaced by random values between 0 m and 3 m or – at around 3% chance – by a signal-not-available value. This randomization has been performed independently for the left and right sides.

The evaluation has been restricted to an urban scenario and the CWUS step with normal belief estimation and likelihood and the mixture model handling each particle independently for each lane-marking observation. While the availability remains around the same for both approaches, the error rate reaches 2.9% for the normal assumption but remains around 0.67% for the mixture model, as shown in Figure 4.18.

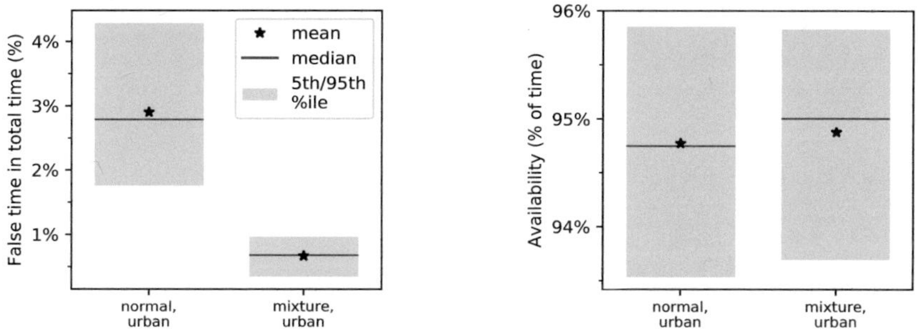

(a) Error rate. **(b)** Availability.

Figure 4.18: Comparison between performance for different combined weight update and sampling settings with corrupted lane-marking data: Normal assumption and mixture assumption with independent modeling. Error rate and availability for 100 repetitions with different random seeds for complete data set and urban scenario.

Overall, the approach that was modeled to allow for outliers within the data does not provide improved performance on the actual sensor data from the vehicle. When introducing artificial noise to the lane-marking measurements, its stability versus erroneous measurements can be seen.

4.2.5.3 Preprocessed Radar Data

Besides the way of handling detected lane-markings, a preprocessing step for radar objects has been proposed in Section 3.3. By using this gating-like method, the localization performance can be improved: With raw radar data, the vehicle is assigned to a wrong lane on average in $r_\mathrm{e} = 0.73\%$ of the time while with preprocessed data, this reduces to the $r_\mathrm{e} = 0.49\%$ found before. The availability is not affected significantly and stays around $r_\mathrm{a} = 96.8\%$, as shown in Figure 4.19. The performance gain is around the same for all three scenarios, i.e. in urban, suburban, and highway environments.

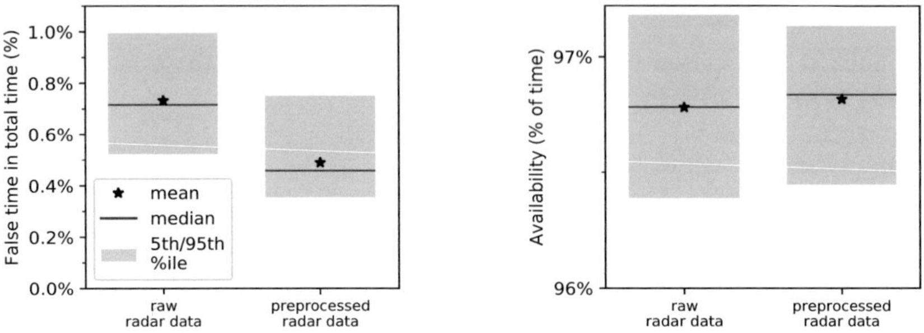

(a) Error rate. (b) Availability.

Figure 4.19: Comparison between performance with and without radar data preprocessing: Error rate and availability for 100 repetitions with different random seeds.

4.2.5.4 Multiple Prediction Models

Incorporating two different yaw rate measurements allows for a gain in the overall localization performance. When using only the ESC yaw rate, the filter reaches $r_\mathrm{e} = 1.3\%$ error rate compared to the $r_\mathrm{e} = 0.49\%$ with both, while the availability stays close to the optimal one at $r_\mathrm{a} = 96.5\%$. In turn, relying only on the gyroscope yaw rate leads to even $r_\mathrm{e} = 4.4\%$ of false lane assignments at an availability of only $r_\mathrm{a} = 93.6\%$. This is mainly due to the gyroscope yaw being calibrated only in standstill and most of the used log files begin with non-zero velocity when entering the mapped area. An overview of these results is shown in Figure 4.20.

4.2.6 Behavior Without Environment Sensors

In the complete sensor setup, three types of data from environment sensors are used: lane-markings, radar objects, and BSM events. To determine the importance of each on the localization performance, the algorithm was run with either of them or combinations being ignored.

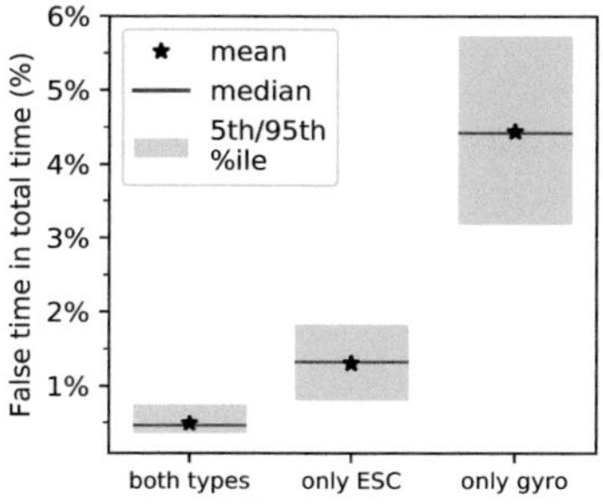

(a) Error rate.

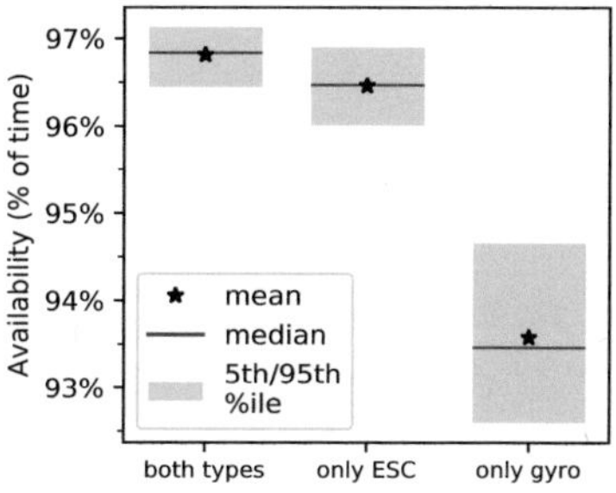

(b) Availability.

Figure 4.20: Comparison between performance with two particle types compared to either type: Error rate and availability for 100 repetitions with different random seeds.

Given the result overview in Figure 4.21, one can see that the influence on the performance on the complete data set differs significantly between the types of data. While disabling BSM with lane-markings and radar objects enabled (*"no BSM"*) has negligible effect ($r_\mathrm{e} = 0.49\%$, $r_\mathrm{a} = 96.5\%$), lane-markings are more crucial to successful lane-assignment ($r_\mathrm{e} = 4.2\%$, $r_\mathrm{a} = 84.5\%$ in *"no lane-markings"*).

There are a few reasons for this: Intuitively, the lane-marking detection only helps to determine the position and orientation within a lane but does not discriminate which lane is used. However, through the combination of vehicle movement in the prediction step and regular measurements on the position and orientation within a lane, lane-changes can be detected in most cases and minor errors in odometry can be compensated. Especially drift in odometry and spreading of the particles over the whole roadbed can be avoided by regular incorporation of lane-marking observations. Like this, it is possible to keep a correct lane-assignment as soon as the filter has converged to the correct hypothesis.

Interestingly, the performance is significantly better without any lane-markings than with incorporating them in traditional weight steps as seen in Section 4.2.5.1. Even when only lane-markings are available and no observation that allows localization relative to other lanes or the road boundary (*"only lane-markings"*), the error rate stays lower ($r_\mathrm{e} = 1.8\%$) at better availability ($r_\mathrm{a} = 91.3\%$). But it cannot reach the performance that can be achieved by a combination of sensors that allow for positioning within a lane and relative to the rest of the road.

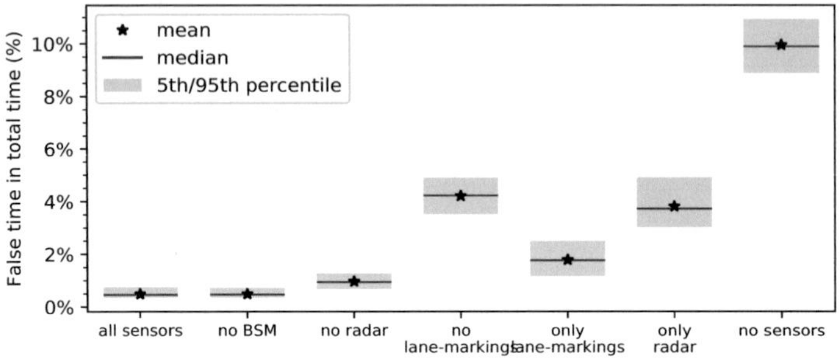

(a) Error rate.

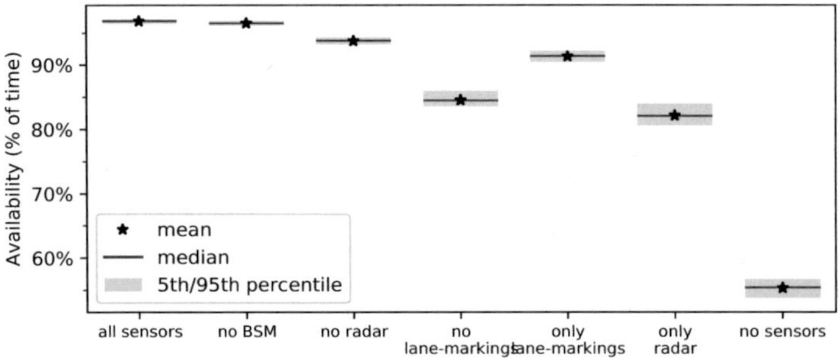

(b) Availability.

Figure 4.21: Comparison of performance with and without certain environment sensors: Error rate and availability for 100 repetitions with different random seeds.

Without radar objects (*"no radar"*), the error rate is around twice the rate with the complete sensor setup ($r_\mathrm{e} = 0.97\%$). Comparing this to the case without radar objects and BSM events (*"only lane-markings"*), one can see that BSM events do actually have an influence on the overall result. This can also be seen when comparing the influence of BSM in the three scenarios (Figure 4.22): In the urban scenario, it seems like using BSM actually increases the error rate from $r_\mathrm{e} = 0.27\%$ to $r_\mathrm{e} = 0.37\%$ and also the variation among the 100 repetitions at a small reduction in availability. The data contains a few situations where the vehicle is e.g. on the rightmost lane but still a BSM event on the right side is triggered. In these situations, using the BSM events as a trustworthy source of information whether the vehicle is on the outermost lane, can be maleficial. In the suburban scenario, however, including BSM increases availability slightly at no increase in error rate. On highways, BSM events are beneficial for the reproducibility as the 95th percentile of the error rate can be reduced to 0.91% compared to 1.5% without BSM. The availability is again slightly increased.

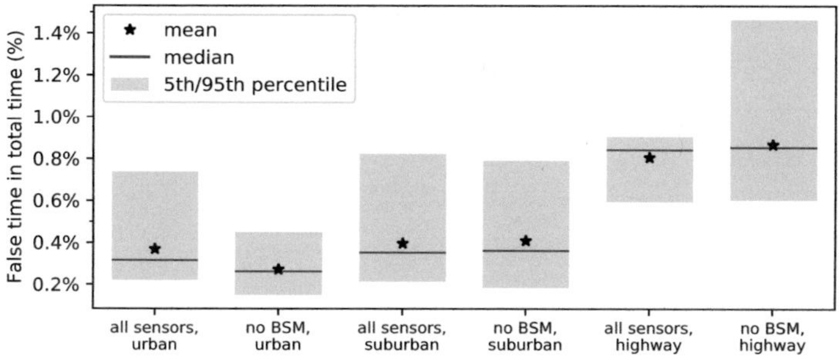

(a) Error rate.

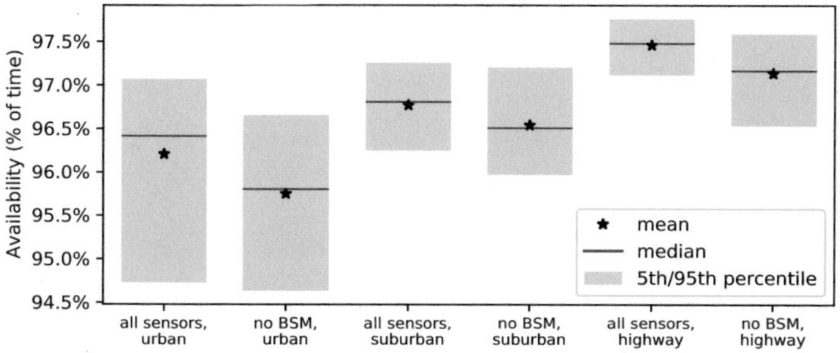

(b) Availability.

Figure 4.22: Comparison of performance with and without BSM in different scenarios: Error rate and availability for 100 repetitions with different random seeds.

Without any information from environment sensors (*"no sensors"*), the availability drops to around $r_{\mathrm{a}} \approx 55\%$ while the error rate increases to $r_{\mathrm{e}} \approx 10\%$. That means, the system is still available and performs a correct lane assignment in around 45% of the time based only on the digital map and vehicle movement. A comparison of the three scenarios shows that for the no sensor-case the error rate is higher than average on highways and the availability is higher than average in suburban scenarios. Both these results are reasonable: On the one hand the map provides less valuable information on highways that would enable the system to benefit from lane-changes and turns. Especially, there are no one-lane sections that have to be passed in order to continue the drive. On the other hand, the suburban scenario has a lower average number of lanes and a noticeable amount of one-lane sections. Within the one-lane section, the filter can assure itself where it is, and within the high number of dual-lane sections, a single lane-change leads to the filter converging to one of the lanes. However, the situation with no environment sensors can be improved by modeling in the particle filter that the vehicle tends to drive in the center of the current lane. A model and evaluation for this is shown in KASTL 2017.

4.3 Particle Filter Localization on Commercial Map

In this section, the performance of the particle filter-based localization on commercial map data, as described in Section 3.5.3 is evaluated. Besides overall results, attention is put on an analysis of situation where the algorithm fails.

4.3.1 Evaluation Metrics

The evaluation is based on the same data as for the version on the lanelet map. However, only five repetitions are performed for each log file and parameter set to get an impression of the achievable error rate and availability. The ground truth is based on the lanelet map ground truth but abstracted from unique lanelet IDs to a lane index of a number of existing lanes. If e.g. the leftmost of three parallel lanelets is correct in the original ground truth, now any lane that is the leftmost of three lanes is considered correct. Furthermore, a slightly adapted evaluation scheme is used. As the current pose estimate may not only be located on mapped lanes but also within intersection polygons (Section 3.5.1.6) or on lanes that are not stored in the map (Section 3.5.1.4), additional result states emerge. Besides the system being not available and the lane estimate being correct or false, these are introduced:

- *intersection*: the majority of the importance weight is on intersections
- *offlink*: the majority of the importance weight is off any mapped road
- *map false*: the number of lanes of the current link estimate does not agree with any correct lanelet group – note that this may either describe a situation where the commercial map has a false lane number attributed or the link hypothesis is wrong
- *map guessed, estimate correct*: the mapped number of lanes of the current link does not agree with the ground truth but the number of lanes including additional lanes (Section 3.5.1.4) does; also the result is correct
- *map guessed, estimate false*: as before, but the result is incorrect

With now eight result states rather than three, a visualization in 2D for different threshold values is not suitable. Hence, charts over different threshold parameters p_{th} are used with the ratios for each of the states stacked. The variations between the minimum and maximum results of every each state are depicted with black error bars.

4.3.2 Quantitative Results

Results are depicted in Figure 4.23. For a threshold of $p_{\mathrm{th}} = 0.7$, the algorithm delivers a false lane estimate on a correct map in 6.2% of the time while a correct estimate can be produced in 67.4% of the time. These values vary among the 15 repetitions with different random seeds between 5.7%–7.7% and 64.7%–68.7%, respectively. In around 7.3% of the time, the system is localized on an intersection and 3.3% of the time off any mapped link. This leaves 3.1% where the map is wrong or the vehicle is localized on a wrong link. In only 0.3% and 0.2% of the time, the number of lanes of the road is guessed and the result is correct or false, respectively. This means that the system is available in 87.7% of the time.

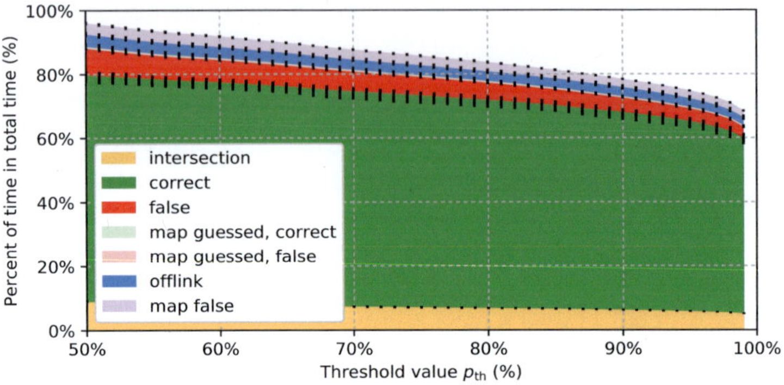

Figure 4.23: Ratios of time for respective result states over threshold parameter p_{th}. The ratios are stacked. Missing percentages to 100% are times where the system is not available. Minimum and maximum ratios of 15 repetitions as black error bars.

4.3.3 Critical Situations

To analyze the challenges of the filtering solution on a current commercial map, some critical situations are analyzed. On one log file with data from Stuttgart, the system provides a false lane assignment in more than 20% of the time. This file contains a couple of situations that are difficult to handle: First, the filter converges correctly to the right lane in a two-lane tunnel segment. Then, the link geometry is inaccurate and does not agree with the dead-reckoned trajectory while at the same time, no lane-markings are detected. The influence of the observed tunnel walls is not high enough, especially it does not change the proposal distribution, to avoid that the particle cloud changes to the left lane. The particle cloud then stays there up to a traffic light after the end of the tunnel. In the following, the particles first follow the actual lane-change to the left and leave the mapped links, but find back to the road and the correct lane through some sections with missing lanes.

After a U-turn, the majority of particles lands on the wrong lane due to a combination of inaccurate longitudinal localization, odometry, and map geometry. While the following lane-change helps the filter to find back to the correct right lane, inaccurate heading estimates before following intersections lead the particles to incorrect lane-changes at following intersections.

Later in the log file, the mapped number of lanes often disagrees with the actual number of lanes. This makes it difficult to find the correct lane in the sections with the correct number of lanes.
Overall, it can be seen that the filter is not very robust against inaccuracies in the map, especially regarding the orientation of link segments, the number of lanes, and unclear lane connections.

4.4 Combination of Particle Filter and Extended Kalman Filter on Commercial Map

The evaluation for the localization scheme with a combination of PF and EKF follows the same structure as for the PF on commercial map data. However, as particles cannot leave the mapped roads, the number of evaluated states reduces from eight to seven. For the evaluation, the number of particles was reduced to $N_{\mathrm{P}} = 400$, as the particle filter state is of lower dimension.

4.4.1 Quantitative Results

While the computational requirements can be reduced with this approach, also the resulting performance decreased. For a fixed threshold parameter $p_{\mathrm{th}} = 0.7$, the algorithm now delivers wrong lane estimates in 11.3% of the time and correct estimates in 53% of the time given correct map data. The vehicle is localized on an intersection in 8.2% of the time and on assumingly wrong map data around 2.9% of the time. The time spent on links where the number of lanes has been guessed correctly is not significant. The variance between different random realizations stays low. An overview of the results for different threshold parameters is depicted in Figure 4.24.

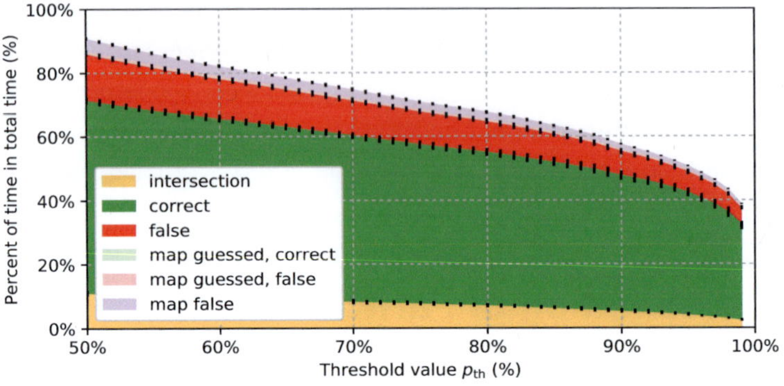

Figure 4.24: Ratios of time for respective result states over threshold parameter p_{th}. The ratios are stacked. Missing percentages to 100% are times where the system is not available. Minimum and maximum ratios of 15 repetitions as black error bars.

4.4.2 Critical Situations

To see situations which are difficult for the combined filtering system, one log file was picked where the performance decreased significantly compared to the pure particle filter approaches and significant situations were analyzed. The filter initially converges onto the correct lane and follows the performed lane-changes. After a turn, it still remains on the correct lane. Then, a lane-change to the right cannot be followed as the map does not yet contain the new lane. The particle filter is reinitialized and follows the next turn but as the heuristic geometry reconstruction connects the correct turn lane to a wrong successor. As the particle filter follows the connected lanes, it moves onto a false lane. The estimated lane remains false as no further lane-change is detected and no relevant radar objects are detected. Only through another reinitialization at a following intersection, where the estimated lane turns, the filter can come back to the correct lane estimate.

From these observations it becomes clear that the filter keeps its lane reliably once converged. But at the same time, it cannot recover from incorrectly connected lanes unless environment sensors provide strong hints on a wrong convergence. It can be seen in other data sets that relying on lane-marking detection only to determine the in-lane position and lane-changes leads to some missed lane-changes.

4.5 Discussion

For navigation systems supporting lane-change recommendations, reliable lane-level localization is needed. While approaches for accurate absolute positioning and localization on landmark-based maps have been subject of publications, using merely a simple geometric map and production vehicle sensors to provide lane-accurate localization in urban, suburban, and highway scenarios has not presented before.

Experimental results show that the solution proposed within this work allows for robust lane-level localization on lanelet maps using a combination of front radar, front camera, inertial sensors, GNSS, and radar-driven BSM available in a current production vehicle. The proposed filter exploits the different types of measurements of this sensor setup. Without the information from the camera, the filter output becomes much less reliable. This shows that the information where the vehicle is located within the lane is important for estimating on which lane the vehicle is.

Surprisingly, BSM deteriorates the performance in urban scenarios while it can improve it in highway scenarios. This is due to the different characteristics of both scenarios: in urban scenarios, BSM provides a couple of false positives where infrastructure has been mistaken for other vehicles. The limitations of this sensor are less grave in a well-structured highway scenario and can support the estimation there.

The proposed combined weight update and sampling method relies on some Gaussian assumptions for the particle distributions which are not completely valid. However, it leads to an improvement in performance compared to a standard particle filter algorithm. These assumptions improve the particle sampling. However, other methods that improve the proposal distribution, such as e.g. an unscented particle filter (VAN DER MERWE 2004), could allow for a similar performance with reasonable amount of particles. Extending the CWUS

approach to mixture distributions of a Gaussian and a uniform do not improve the performance on the given data set.
Through the evaluations, it also becomes clear that the proposed solution relies heavily on accurately mapped road and lane geometry. Despite several adaptations that model the uncertainty of a current commercial navigation map, the achieved performance is significantly worse with higher error rates and lower availability. A combination of a particle filter and Extended Kalman Filter was proposed to better handle the geometric inaccuracy of the map. But while strongly restricting the particles to their respective current lane avoids erratic behavior in some situations, the result did not improve over the pure particle filter approach. Instead, its inability of free movement even increases the error rate further.
This leads to the conclusion that accurate lane-geometry in the digital map is beneficial for reliable lane-level localization. It can be expected that similar results to those on the lanelet map can be achieved on commercial lane-level maps such as HERE HD Live Map (BONETTI 2016) or TomTom Highly Detailed Map (TOMTOM 2016). To further improve the results, hints supporting the longitudinal localization of the vehicle could help, such as mapped pedestrian crossings or stop lines.
Despite the extensive data set used for the evaluation, the system has only been tested in a restricted geographic area. Before worldwide use, it should be evaluated in multiple countries with different traffic situations and types of road markings.

5 Application Example: Automotive AR Navigation

Route guidance is one of the most promising use cases of automotive AR systems. After an overview of related work in Augmented Reality in Section 5.1, a system architecture is defined for such a system based on the localization schemes in Chapter 3. Further issues related to localization for this application are treated in Sections 5.3 and 5.4.

5.1 Related Work

AR applications can be divided into image-based and position-based approaches. The first group adds virtual objects to a local environment whose shape is known or modeled. The approaches to localization include tracking features or landmarks in the field-of-view of a camera and are similar to visual odometry (SCHÖPS et al. 2014) and SLAM (SALAS-MORENO et al. 2014) methods known from feature-based localization (Section 3.1). These methods are often run on handheld devices such as smartphones and require low computational effort (SALARIAN and ANSARI 2015).

In the other group, one tries to accurately determine the global position and display virtual objects whose positions are known in absolute coordinates. As for automotive route guidance, a global position and/or a location relative to a digital map – which in turn is given in global coordinates – is required, the focus in this overview lies on position-based AR and AR navigation applications.

Methods for estimating the 2D-pose are proposed by PARK et al. 2012 and LIN et al. 2011. PARK et al. improve the pose estimate by using edges from camera data and buildings in a digital map for location based services. Panoramic images are employed for tracking by LIN et al. to display points of interest (POIs) in automotive AR. However, as the orientation seems to derive from the position differences between two consecutive GPS fixes at 1 Hz and no extrapolation is described, the estimated pose is probably not available at high-frequency. Furthermore, the 3D-pose estimate including all three orientation angles is required for proper overlay of virtual objects over real environment.

SCHALL et al. 2009 describes a method to estimate this 3D-pose for handheld devices: DGPS/RTK data is fused with barometric heights, information from an IMU, and a visual orientation tracker. Another approach using an EKF to estimate the current pose in a tightly coupled IMU, GPS and image data scenario is proposed in OSKIPER et al. 2012. It is used for outdoor position based AR with head-mounted displays (HMDs).

Among the first adopters to AR route guidance are HU and UCHIMURA. For their vision-based car navigation system (VICNAS), they estimate the absolute vehicle pose based on DGPS and gyroscope data in a simple extrapolation from the last trustable GPS position. The pose estimate is supported by a model matching between features in image data and road shape data from a digital map. A prototypical user interface allows for navigation by superposition of virtual objects on the camera image (Figure 5.1a) (HU and UCHIMURA 2004). The approach has later been extended to estimate the road geometry in front of the vehicle. In an EKF, parameters such as the width of the lane, the lateral displacement to the lane-marking, clothoid parameters, and the camera orientation are determined based on map and camera data (WANG et al. 2011). A resulting road geometry estimate is shown in Figure 5.1b.

(a) Maneuver-based. **(b)** Lane-geometry.

Figure 5.1: Implementations of AR navigation system VICNAS. Images taken from HU and UCHIMURA 2004 and WANG et al. 2011, respectively.

Further AR navigation systems are demonstrated by the University of Linz and Siemens AG (SIEMENS AG 2005) as well as by the University of Karlsruhe and Harman Becker Automotive Systems (SHEN 2013; VACEK et al. 2006). VACEK et al. track multiple lanes using particle filters and detect intersections by finding intersecting lanes. Supported by navigation system data, direction attributes are assigned to the found lanes and arrows are drawn on recommended lanes. As seen in Figure 5.2a, the system works well in example situations but the authors note that the intersection extraction and lane attribution are not yet robust enough and only three lanes are supported.

(a) VACEK et al. 2006.

(b) LEE et al. 2015.

Figure 5.2: Implementations of AR navigation system in VACEK et al. 2006 (left) and LEE et al. 2015 (right). Images taken from respective publications.

Another lane-level AR navigation system is proposed in LEE et al. 2015. Lane-change recommendations are visualized in a head-up display (HUD) (Figure 5.2b) but the system is only evaluated in an indoor testbed. No details are given on how the currently used and recommended lanes are determined.
Further insight on the challenges and experiences during the development of an automotive AR system is given by RAO et al. The influence on the E/E architecture of a vehicle regarding bandwidth, latency, and synchronization are highlighted in RAO et al. 2014a while the overall system design, pose estimation algorithm, and test setup are described in RAO et al. 2014b.
A further interesting use case besides route guidance and lane-marking highlighting is shown by SCHÜLE: Given a relative localization on a digital road map using radar and camera data, the upcoming road geometry is highlighted in a night vision camera image to provide guidance on rural roads (Figure 5.3a).

(a) SCHÜLE et al. 2013.

(b) GEORGE et al. 2012.

Figure 5.3: AR road course information (left) and metaphors for obstacle detection system (right). Images taken from respective publications.

Other research puts more focus on human-machine interface (HMI) aspects. BERGMEIER and BUBB 2008 compare different HMI variants in a non-stereoscopic AR-HUD with regard to the perceived depth. Use cases are navigation, driver assistance systems, and night vision. A detailed study on use cases and concepts for automotive AR is also given by TÖNNIS 2008. With user studies in a prototypical implementation in a driving simulator, questions like the creation of spatial context and guiding the attention in dangerous situations are discussed. A visualization metaphor for driver assistance systems is developed in GEORGE et al. 2012. The system monitors what the driver is looking at and only highlights what is detected by a Mobileye camera but unseen by the driver. The visualization shown in Figure 5.3b displays different types of danger, levels of dangerousness and criticality in a single metaphor.

5.2 System Architecture

The lane-precise AR navigation system can be divided into three main parts as depicted in Figure 5.4: localization, routing, and visualization. The localization block determines the currently used road link or lane. For the case that no lane-precise localization is possible, a heuristic approach based on a map-matching result is employed. These methods are described in detail and evaluated in the following sections. Localization logic decides on a suitable localization result and provides the respective information to other components.

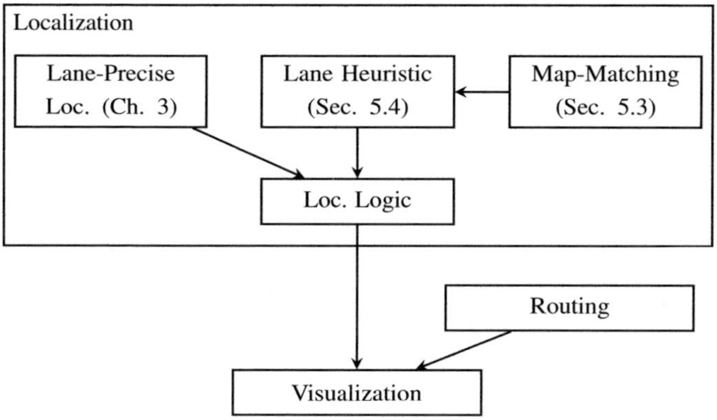

Figure 5.4: System architecture of AR navigation.

Routing takes as an input the desired destination and calculates a suitable route consisting of links and recommended lanes leading to that destination. Existing lane-level navigation systems mostly recommend one or multiple lanes at a maneuver. For a route guidance system that provides lane-change recommendations, the routing block must also produce lane recommendations for road segments when approaching a maneuver. For this purpose, the road network graph needs to be traversed against travel direction up to the current link, if possible.

The visualization section differs significantly from traditional navigation systems: First, it determines based on localization and routing, whether a lane-change is required within the upcoming road section. Then, a navigation carpet is generated using the digital map and sensor data. The resulting carpet is then sent to the rendering engine to be displayed in an AR video.

For the exemplary results shown in Figure 5.5, the lane-precise localization described in Section 3.4 is used whenever possible and complemented by heuristic lane-change recommendations (Section 5.4). Routing and visualization are implemented in a simple manner: For turn right and exit right maneuvers, the vehicle is guided as far right as possible in steps of one lane; similarly for turn left maneuvers. For the visualization, a navigation carpet is generated directly from commercial map data and moved laterally such that the closest part of the navigation carpet lies on the ego-lane or one of the neighboring lanes.

Figure 5.5: Examples for AR lane-change recommendation. Ego-lane highlighted in yellow in first image.

5.3 Map-Matching for Automotive AR

*This section is based on the work of Andreas Danner during his internship (*DANNER *2015: implementation, parametrization) and the resulting publication (*RABE *et al. 2015).*

Traditional navigation systems employ a map-matching step to determine the currently used road. In this step, visualized in Figure 5.6a, the current position estimate, for example a GNSS fix, is pulled to the closest point on a mapped road. In modern variants, not only a single point is considered but also trajectories or sets of points and map topology (AUER et al. 2017; GEORGY et al. 2012; QUDDUS and VELAGA 2012). In both cases, however, the road is assumed one-dimensional. This leads to a problem when using this kind of map-matching in an AR navigation system, depicted in Figure 5.6b: When a trajectory such as the blue one is map-matched, this is likely to result in the red dashed line. Even if the angular components of the matched trajectory are kept free, this will lead to virtual objects in a video image moving around rather than staying at virtually fixed position. As soon as the trajectory starts into the turn, the vehicle and its AR frame move to the right while the map-matched position goes straight ahead.

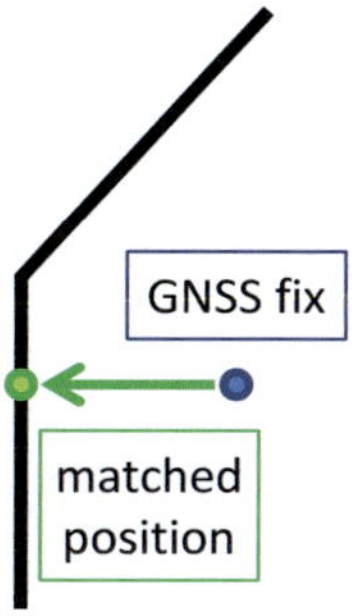

(a) Map-matching.

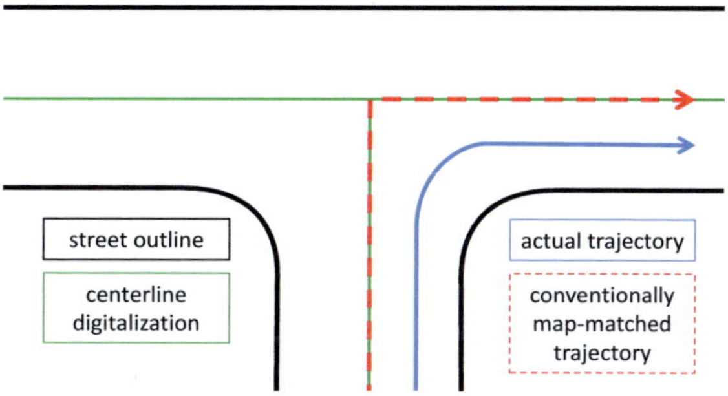

(b) Problem with conventional map-matching in AR.

Figure 5.6: Principle of conventional map-matching and challenge in case of AR system (RABE et al. 2015).

In this case, virtual objects seem to move to the right with the vehicle. At some point during the turn, the map-matching will switch from the link leading to the intersection to the new link. At this moment, all virtual objects will instantly change their appearing position.

On the other hand, an AR system that displays objects with absolute coordinate reference stored in a map requires some kind of map-matching. As the estimated pose is based on GNSS measurements with a local Dead Reckoning (RAO et al. 2014b), it might exhibit a noticeable offset to its actual position. If this position was used directly in an AR system, virtual objects, such as maneuver arrows at intersections or house numbers, would appear with an offset as well. Hence, a map-matching scheme is required that allows for some free movement to allow realistic movement of virtual objects but also constrains the position to the map to allow for accurate object positioning. A possible method is shown in overview in Figure 5.7 and explained in the following.

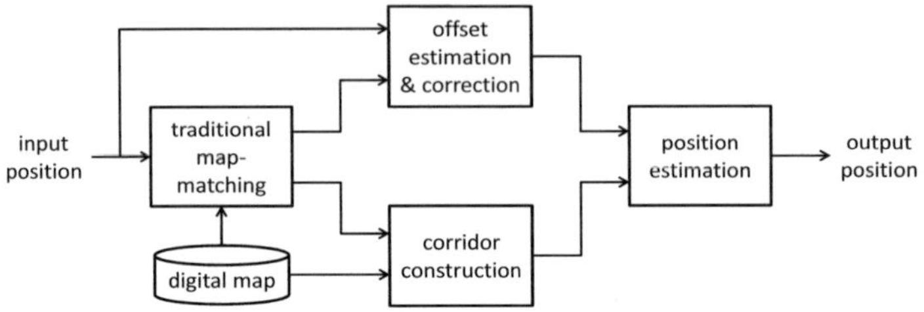

Figure 5.7: Block diagram of proposed algorithm (RABE et al. 2015).

5.3.1 Traditional Map-Matching

First, the closest link to the pose estimate $\boldsymbol{x}_{\mathrm{i},k}$ is determined, taking into account the distance and difference in heading between the pose and the link's centerline. The previously matched road segment is considered to avoid continuous road switching in case of adjacent roads. This block yields the current road segment and the vehicle's position matched to the road centerline $\boldsymbol{x}_{\mathrm{mm},k}$.

5.3.2 Allowed Driving Corridor

The corridor construction block determines a function that maps an absolute input distance to the centerline d_{in} to a similar output distance. The transfer function is designed such that lane-changes and obstacle avoidance maneuvers – i.e. the vehicle's small-signal behavior – are mapped directly while the total deviation to the lane centerline is limited to reasonable values. A possible implementation for one-way roads is (RABE et al. 2015)

$$g(x) = \begin{cases} \tanh(3x - 1.3014), & x > 0.8 \\ x, & x \leq 0.8, \end{cases} \tag{5.1}$$

where $x = \frac{d_{\mathrm{in}}}{w}$ for the estimated road width $w = 3\,\mathrm{m} \cdot n_{\mathrm{lanes}}$.
This design allows for free movement in the inner 80% of the road and leads the output position deviation into saturation when moving toward the edge of the road. The numeric parameters are chosen to allow a smooth transition between both parts. If desired, the inner area can be widened by changing the threshold;

the outer area can allow for larger deviations by reducing the factor in the $\tanh()$ function and adapting its shift accordingly.
As roads may widen up with additional lanes when approaching intersections, a wider corridor is used in these cases. Similarly, a comparable transfer function is used for two-way roads.

5.3.3 Offset Vector Estimation and Correction

The small-signal behavior of the relative pose is represented optimally when the input pose stays within the corridor defined in Equation (5.1). To achieve this even for larger GNSS offsets, an offset vector is estimated based on the input and map-matched positions. As a point-to-line-matching only allows for estimation of the lateral deviation – and effect also observed in SCHREIBER et al. 2013 when matching against line features in a camera image –, separate offset vectors are stored for different vehicle headings: The possible headings are split into 24 sectors of $15°$ and the offset vector for any direction is only estimated when traveling almost ($\pm 7.5°$) in the respective orientation. The resulting observed offset vector $\boldsymbol{o}_{s,k}$ for sector s is always perpendicular to the respective sector orientation and defined as the exponentially filtered difference in x- and y-components between the input position $\boldsymbol{x}_{\mathrm{i},k}$ and the map-matched position $\boldsymbol{x}_{\mathrm{mm},k}$,

$$\boldsymbol{o}_{s,k} = \begin{bmatrix} o_{x,s,k} \\ o_{y,s,k} \end{bmatrix} = (1-\alpha) \begin{bmatrix} o_{x,s,k-1} \\ o_{y,s,k-1} \end{bmatrix} + \alpha \begin{bmatrix} x_{\mathrm{mm},k} - x_{\mathrm{i},k} \\ y_{\mathrm{mm},k} - y_{\mathrm{i},k} \end{bmatrix}, \tag{5.2}$$

where α is set to a value around $\alpha \approx 10^{-3}$ to accommodate for the slow changes in the estimated offset at a pose frequency of 100 Hz.
The longest of the estimated offset vectors $\boldsymbol{o}_k = \max_s \|\boldsymbol{o}_{s,k}\|$ corresponds to the actual offset and is used for correcting the input position before the actual AR map-matching step. To reduce the influence of old estimates that might have become invalid, each offset vector is only considered valid for a certain distance after its last change. Influence of outliers is reduced by only applying the estimated offset when at least three offset vectors are valid.

5.3.4 Position Estimation

The actual map-matching step consists of solely applying the transfer function from Equation (5.1) onto the offset corrected position estimate

$$\begin{bmatrix} x_{\mathrm{corr},k} \\ y_{\mathrm{corr},k} \end{bmatrix} = \begin{bmatrix} o_{x,k} \\ o_{y,k} \end{bmatrix} + \begin{bmatrix} x_{\mathrm{i},k} \\ y_{\mathrm{i},k} \end{bmatrix}. \tag{5.3}$$

The map-matched position estimate can then be used by the AR engine.

5.3.5 Results

The proposed scheme can be evaluated by observing resulting trajectories superimposed on satellite images and in AR videos. Position data from around one hour of drives around Stuttgart have been visually evaluated using Google Earth. Figure 5.8a depicts how the input trajectory (green) is matched onto the road (purple trajectory): While the general shape is preserved and hence virtual objects keep their respective positions, the trajectory is moved onto the mapped road link and a possible offset of absolutely positioned objects is reduced.

(a) Urban intersection.

(b) Tunnel exit (left).

Figure 5.8: Exemplary results: input trajectory in green, map-matched trajectory in purple (Rabe et al. 2015). Image source: GeoBasis-DE/BKG, Google Earth.

When traveling through tunnels, the pose estimate may drift due to the lack of GNSS reception. At the end of the tunnel, new GNSS fixes are acquired and the estimated pose is corrected accordingly. This effect is shown in Figure 5.8b where a tunnel on the left is left and the input trajectory (green) is corrected.

(a) Straight arrow. **(b)** Turn right arrow.

Figure 5.9: Position of objects on map. Travel direction from top right to bottom left (RABE et al. 2015). Image source: GeoBasis-DE/BKG, Google Earth.

The map-matched trajectory (purple) stays close to the mapped link throughout the tunnel and even stays within the roadbed when the input is corrected.
To evaluate the influence on offset reduction of virtual objects with absolute coordinates, two such objects have been created in their assumed perfect position: A straight arrow on the middle of three lanes (Figure 5.9a) and a turn right arrow on the rightmost lane (Figure 5.9b). The situations are evaluated in an urban scenario with a significant GNSS offset of more than 10 m.

(a) Without map-matching. **(b)** With map-matching.

Figure 5.10: Straight arrow in video (RABE et al. 2015).

In Figure 5.10a, one can see that the straight arrow seems to be positioned on the physical divider between the directional roadbeds rather than on the middle lane. Comparison with the longitudinal position with map-matching (Figure 5.10b) shows that it actually positioned on the opposing roadbed behind the physical divider. It is only interpreted as being positioned within the physical divider due to the lack of occlusion of virtual objects behind real objects. The recognized

position in the image with map-matching, however, lies within the central lane as intended by the absolute position.
The same effect occurs with the turn right arrow depicted in Figure 5.11. Again, an offset to the left can be seen when using the input position. With the proposed map-matching scheme, the arrow is positioned within the rightmost lane as intended.

(a) Without map-matching.

(b) With map-matching.

Figure 5.11: Turn right arrow in video (RABE et al. 2015).

The proposed map-matching scheme leads to the desired results when analyzed in satellite images and AR video. While the offset of virtual objects with absolute positions is reduced, the shape of the original trajectory is kept and map-matching does not introduce noticeable movement of stationary virtual objects in AR video.

5.4 Lane-Change Recommendations in Unclear Situations

*This section is based on the work of Benjamin Joswig during his internship (*JOSWIG *2016: implementation, parametrization) and the resulting publication (*RABE *and* JOSWIG *2016).*
Current commercial infotainment maps contain the complete number of lanes including turn lanes mainly in front of large urban intersections and highway interchanges. In other situations, the number of through lanes is attributed to provide an estimate for the importance and throughput of the road. Lane-connectivities are not known in those situations. However, lane-level navigation

may also be desired in these cases where it is not possible to determine recommended lanes and perform a reliable ego-lane assignment. In these cases, a different approach to lane-level navigation is required that combines lane-level localization and lane recommendation. Its goal is to determine whether another lane exists in the direction of the next maneuver, e.g. whether the ego-vehicle is positioned on the rightmost lane when approaching a turn right maneuver. This differs from many localization problems in that the system should not find out where the vehicle is located but rather where it is not. These situations are called *possible lane-changes* in the following.
This approach is based on vehicles detected by the front radar and BSM system and lane-marking detection. It further incorporates vehicle speed and turn indicator state for its heuristic.

5.4.1 Determining Possible Lane-Changes

Turn maneuvers are provided by the system 100 m to 500 m before the actual maneuver. Within this section, the system can determine possible lane-changes. Whenever a turn maneuver is provided, the algorithm determines whether another intersection lies on the way to the maneuver. In that case, it could only find turn lanes for the upcoming intersection and not for the maneuver intersection. Hence, it waits until the next intersection and then looks for further intersections. If no more intersection lies between vehicle and maneuver, the system uses the data from the environment sensors to estimate separate scores. These scores relate to the probability of the existence of another lane in turn direction, i.e. for possible lane-changes.
For the lane-marking-based score, highway-like lane-markings are assumed, i.e. the outermost lane-markings are continuous, the inner ones some kind of broken type. The score is increased every time a non-continuous lane-marking is detected in maneuver direction. The increase depends on the number of mapped lanes at the current position and the detected lane-marking types in both directions as depicted in Figure 5.12. The motivation for this is that before a turn right maneuver, a continuous left lane-marking gives a strong hint on being on the leftmost lane, leading to an increased chance for further lanes to

the right. Accordingly, the number of lanes according to the map – which might be zero if no lane-attribution is available – gives an idea of the size of the road.

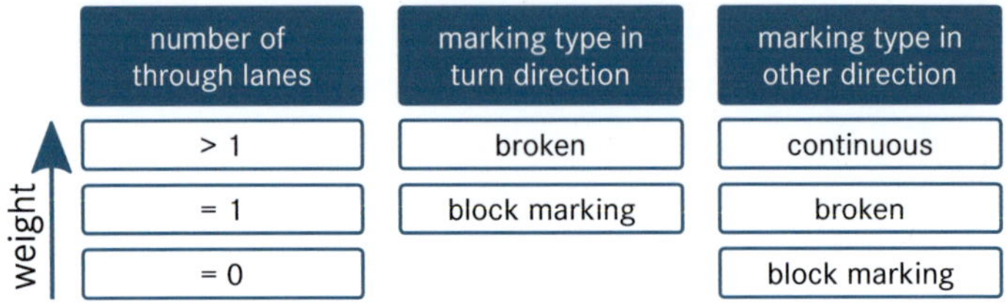

Figure 5.12: Influence of detected lane-markings on score.

The precise values for the score weights were tuned empirically in sample situations. The score is automatically reset if a continuous lane-marking in turn direction has been detected at least five times. For turn left maneuvers on bidirectional roads, this data is not used as the central lane separator to the oncoming traffic may be mistaken for a lane-marking between adjacent lanes.
Similar scores are determined based on data from the front and blind spot radars: Whenever an object has been detected in front of the vehicle and assigned to the same relative lane multiple times, it is considered valid. A lane-change in maneuver direction is assumed possible as soon as one radar object is considered valid. For the BSM system, a single vehicle in the blind spot leads to maximum score.
Whenever a lane-change is performed, it needs to be reconsidered whether another lane-change is possible. Hence, the system tries to determine lane-changes from the available sensor data. Of course, lane-changes observed by the lane-marking detection are considered. However, further methods are required as the system does not always provide lane-changes and lane-markings are not always visible. Whenever the relative lanes of two valid radar objects change within a close time period, it can be assumed that a lane-change has been performed. Alternatively, a lane-change of one valid object during an active turn indicator also gives a hint on a performed lane-change.

5.4.2 Results

The heuristic has been evaluated on 144 turn maneuvers in urban and highway situations. Of these, 85 were turn right maneuvers, 59 turn left maneuvers. Possible lane-changes are denoted as *positives* and situations where the vehicle is already on the outermost lane as *negatives*. It is crucial to avoid false positives as they might lead the driver onto grass verges, sidewalks, or into oncoming traffic. Hence, the system has been tuned to rather not recommend anything than prematurely recommending lane-changes.

The results are shown in overview in Table 5.1. It can be seen that a majority of possible lane-changes can be detected for turn right maneuvers. In these cases, the possible lane-changes are detected after 2.8 seconds or 20 meters, on average. The false positives occur in situations with very short turn lanes or with intersections shortly before the actual maneuver. The one false positive occurs in a situation where a moving object is detected to the right of the ego-vehicle, i.e. on the sidewalk.

Table 5.1: Detected possible lane-changes (RABE and JOSWIG 2016).

	Turn right		Turn left	
	Negative	Positive	Negative	Positive
False	15.3%	1.2%	18.6%	1.7%
True	31.8%	51.8%	42.4%	37.3%
Precision	97.8%		95.7%	
Recall	77.2%		66.7%	

In two thirds of the possible lane-changes to the left, this was detected. The increased false negative ratio, compared to the turn right scenario, is due to the ignorance of the lane-marking detection on bidirectional roads. Again, one false positive occurs. This time, it is due to a curb being continuously classified as a broken lane-marking. For the true positives, possible lane-changes were found after, on average, 3.9 seconds or 25 m.

The evaluation has been performed offline on the same data for turn left and turn right maneuvers. Therefore, the vehicle rarely followed recommended lane-changes. Six lane-changes have been performed during turn right maneuvers; three before and three after a possible lane-changes had been detected. All of them could be detected by the lane-change recognition. No lane-changes to the left occurred during active turn left maneuvers.

It can be seen that the method can support a traditional localization scheme in situations where the number of total lanes is unsure. With only few false positives, the method can be used for an infotainment system. As the data from the three sensors are used completely independent, it can also be used when not all sensors are available.

6 Conclusion and Outlook

This work showed that it is possible to reliably determine the currently driven lane using only current production vehicle sensors and a digital map containing lane-geometry data. Neither highly accurate global positioning using DGPS or RTK-GPS nor feature- or point-landmark-based maps are required. The developed method enables next-generation automotive route guidance applications such as lane-level AR navigation and lane-change recommendations in traditional navigation systems.

In Chapter 1, this work was motivated and its main contributions described. Chapter 2 provided an insight into fundamentals and restrictions of the used map and sensor data and gave a short introduction into dead reckoning and Bayes estimation.

The localization algorithm was developed in Chapter 3: Two data preprocessing steps are performed before the actual localization filter. The localization is based on a particle filter approach that is extended by a combined weight update and sampling step to allow the use of tight likelihoods even when the particles are spread widely over the state space.

The approach is adapted to current commercial infotainment maps in two ways. For both, map data is preprocessed and some changes to the handling of sensor and map data are made. One version is also based on a particle filter with a combined weight update and sampling step, the other combines an Extended Kalman Filter and a particle filter over the remaining dimensions.

Chapter 4 assesses the results achieved with the proposed approach: A simulation showed that the reproducibility of the results is increased by the use of the proposed combined weight update and sampling step in both the Gaussian and mixture variant, compared to a traditional importance weighting step on lanelet maps. Furthermore, smaller map errors can be handled better when employing a combined weight update and sampling step.

Evaluations on log data from real drives in urban, suburban, and highway scenarios showed that error rates of 0.5% can be achieved with the system being available in more than 96% of the time. In a comparison between the combined weight update and sampling step and traditional importance weighting, the benefits of this step could be shown again. However, the extension of using mixture distributions in the combined weight update and sampling rather than simplifying it to Gaussians does not bring a significant advantage on the used data set. Unavailability of sensors like visual lane-marking detection and radar objects reduces the achievable performance just like using only one type of odometry rather than two different sources.
The evaluation of the localization adapted to commercial map data showed that in some situations, the filter can still find the currently used lane. However, the overall availability and accuracy are reduced.
Chapter 5 puts the proposed localization solution into the context of an augmented reality navigation system: First, a possible system architecture that allows for separate design and analysis of individual components of the navigation system is described. Then, two algorithms that can be useful for tasks required in an AR navigation system are proposed and briefly evaluated. One allows for map-matching on a traditional commercial navigation map by defining a corridor in which the vehicle may move and softly restricting the estimated position within this corridor. The other uses available sensor and map data in a heuristic manner to determine whether a drivable lane is available in the turn direction at an upcoming navigation maneuver. This allows for a lane-level navigation in situations where precise map data is not available or the algorithm presented in Chapter 3 cannot provide a certain lane estimate.
Due to the uncertainty of the localization on commercial map data, further development in this direction might be needed or, alternatively, map data with accuracy comparable to the used lanelet maps is required. In this context, further adaptations can be thought of: To avoid losing the correct hypothesis, it can be useful to switch from a particle filter method to a 3D histogram filter around the current GNSS position or around a map-matched position from another system. Instead of combining an EKF and PF, a discrete Bayes filter ("histogram filter") can be used to distinguish between a set of in-lane positions determined with an EKF. The respective EKFs can be reinitialized after intersections to exploit

the acquired information on the lateral (then longitudinal) position and to avoid relying on possibly uncertain longitudinal (then lateral) position information. To avoid errors through map inaccuracies, the map can be modeled probabilistically as in SLAM methods. To this end, information such as the number of lanes and the connection between lanes of different links can be modeled as a part of each particle's state.

Considering the lane-level AR navigation, further work on visualization, i.e. the fusion of map and sensor data on the road in front of the vehicle, is required. Together with lane-level routing and the presented localization method, important steps toward a intuitive new generation of automotive navigation systems can be taken.

References

ALTHOFF et al. 2017
M. Althoff, M. Koschi, and S. Manzinger (2017). "CommonRoad: Composable Benchmarks for Motion Planning on Roads." In: *IEEE Intelligent Vehicles Symp. (IV)*. Redondo Beach, USA, June 2017, pp. 719–726

ARULAMPALAM et al. 2002
M. S. Arulampalam, S. Maskell, N. Gordon, and T. Clapp (2002). "A tutorial on particle filters for online nonlinear/non-Gaussian Bayesian tracking." In: *IEEE Trans. Signal Processing* 50.2, pp. 174–188

AUER et al. 2017
M. Auer, H. Rehborn, S.-E. Molzahn, and K. Bogenberger (2017). "Boosting Performance of Map Matching Algorithms by Parallelization on Graphics Processors." In: *IEEE Intelligent Vehicles Symp. (IV)*. Redondo Beach, USA, June 2017, 462–467

AZUMA 1997
R. T. Azuma (1997). "A survey of augmented reality." In: *Presence: Teleoperators and virtual environments* 6.4, pp. 355–385

BAUER et al. 2016
S. Bauer, Y. Alkhorshid, and M. Wanielik (2016). "Using High-Definition Maps for Precise Urban Vehicle Localization." In: *19th Int. IEEE Conf. on Intelligent Transportation Systems (ITSC)*. Rio de Janeiro, Brazil, Nov. 2016, pp. 492–497

BENDER et al. 2014
P. Bender, J. Ziegler, and C. Stiller (2014). "Lanelets: Efficient Map Representation for Autonomous Driving." In: *IEEE Intelligent Vehicles Symp. (IV)*. Dearborn, Michigan, USA, June 9, 2014

BERGMEIER and BUBB 2008
U. Bergmeier and H. Bubb (2008). "Augmented Reality in Vehicle – Technical Realisation of a Contact Analogue Head-Up Display Under Automotive Capable Aspects: Usefulness Exemplified through Night Vision Systems." In: *Proceedings of 32nd World Automotive Congress (FISITA)*

BETAILLE and TOLEDO-MOREO 2010
D. Betaille and R. Toledo-Moreo (2010). "Creating Enhanced Maps for Lane-Level Vehicle Navigation." In: *IEEE Trans. Intell. Transp. Syst.* 11.4 (Dec. 2010), 786–798. ISSN: 1524-9050, 1558-0016

BIENVENÜE et al. 2002
A. Bienvenüe, M. Joannides, J. Bérard, É. Fontenas, and O. François (2002). "Niching in Monte Carlo Filtering Algorithms." In: *Artificial Evolution: 5th International Conference, Evolution Artificielle, EA 2001 Le Creusot, France, October 29–31, 2001 Selected Papers*. Ed. by P. Collet, C. Fonlupt, J.-K. Hao, E. Lutton, and M. Schoenauer. Berlin, Heidelberg: Springer Berlin Heidelberg, pp. 19–30. ISBN: 978-3-540-46033-6

BMW GROUP 2017
BMW Group (2017). *BMW i erprobt „Augmented Reality Product-Visualiser" mit Tango, der Smartphone-AR-Technologie von Google*. May 1, 2017. URL: https://www.press.bmwgroup.com/deutschland/article/detail/T0266825DE/bmw-i-erprobt-%E2%80%9Eaugmented-reality-product-visualiser%E2%80%9D-mit-tango-der-smartphone-ar-technologie-von-google?language=de (visited on 08/30/2017)

BONETTI 2016
P. Bonetti (2016). *HERE introduces HD Live Map to show the path to highly automated driving*. Jan. 5, 2016. URL: http://360.here.com/2016/01/05/here-introduces-hd-live-map-to-show-the-path-to-highly-automated-driving/ (visited on 08/17/2016)

CARON et al. 2007

F. Caron, M. Davy, E. Duflos, and P. Vanheeghe (2007). "Particle filtering for multisensor data fusion with switching observation models: Application to land vehicle positioning." In: *Signal Processing, IEEE Transactions on* 55.6, pp. 2703–2719

CHARMETTE et al. 2009

B. Charmette, E. Royer, and F. Chausse (2009). "Matching planar features for robot localization." In: *Advances in Visual Computing*. Springer, pp. 201–210

CHAUSSE et al. 2005

F. Chausse, J. Laneurit, and R. Chapuis (2005). "Vehicle localization on a digital map using particles filtering." In: *IEEE Intelligent Vehicles Symp. (IV)*. Las Vegas, USA, June 2005, pp. 243–248

CHEN 2003

Z. Chen (2003). "Bayesian filtering: From Kalman filters to particle filters, and beyond." In: *Statistics* 182.1, pp. 1–69

CORETTO and HENNIG 2011

P. Coretto and C. Hennig (2011). "Maximum likelihood estimation of heterogeneous mixtures of Gaussian and uniform distributions." In: *Journal of Statistical Planning and Inference* 141.1, pp. 462–473

CORNICK et al. 2016

M. Cornick, J. Koechling, B. Stanley, and B. Zhang (2016). "Localizing Ground Penetrating RADAR: A Step Toward Robust Autonomous Ground Vehicle Localization." In: *Journal of Field Robotics* 33.1, pp. 82–102. ISSN: 1556-4967

DAIMLER 2012

Daimler AG (2012). *Mercedes-Benz S-Class, technology, radar, Stereo Multi-Purpose Camera (SMPC, stereo camera for short), ultrasonic sensors*. Nov. 19, 2012. URL: `http://media.daimler.com/marsMediaSite/en/instance/picture.xhtml?oid=7599386` (visited on 10/25/2016)

DANNER 2015
A. Danner (2015). *Anforderungsspezifische Map-Matching Erweiterung für ein Augmented Reality System im Fahrzeug*. Praktikumsbericht. Sindelfingen: DHBW, June 2015

DAO et al. 2006
T.-S. Dao, K. Y. K. Leung, C. M. Clark, and J. P. Huissoon (2006). "Co-operative lane-level positioning using Markov localization." In: *9th Int. IEEE Conf. on Intelligent Transportation Systems (ITSC)*. Toronto, Canada, Sept. 2006, 1006–1011

DAO et al. 2007
— (2007). "Markov-based lane positioning using intervehicle communication." In: *IEEE Trans. Intell. Transp. Syst.* 8.4, pp. 641–650

DOUC and CAPPÉ 2005
R. Douc and O. Cappé (2005). "Comparison of resampling schemes for particle filtering." In: *4th Int. Symp. on Image and Signal Processing and Analysis (ISPA)*, pp. 64–69

DOUCET et al. 2001
A. Doucet, N. De Freitas, and N. Gordon (2001). *Sequential Monte Carlo methods in practice*. Statistics for Engineering and Information Science. Springer

DOUCET et al. 2000
A. Doucet, S. Godsill, and C. Andrieu (2000). "On sequential Monte Carlo sampling methods for Bayesian filtering." In: *Statistics and computing* 10.3, pp. 197–208

DOUCET and JOHANSEN 2009
A. Doucet and A. M. Johansen (2009). "A tutorial on particle filtering and smoothing: Fifteen years later." In: *Handbook of Nonlinear Filtering* 12, 656–704

DU and BARTH 2008
J. Du and M. J. Barth (2008). "Next-generation automated vehicle location systems: Positioning at the lane level." In: *IEEE Trans. Intell. Transp. Syst.* 9.1, pp. 48–57

DU et al. 2004
J. Du, J. Masters, and M. Barth (2004). "Lane-level positioning for in-vehicle navigation and automated vehicle location (AVL) systems." In: *7th Int. IEEE Conf. on Intelligent Transportation Systems (ITSC)*. Washington, D.C., USA, Oct. 2004, pp. 35–40

ECKEL 2015
I. Eckel (2015). "Particle Filters for Airborne Tracking and Lane-Level Map-Matching of Vehicles." PhD thesis. München: Technische Universität München

EL NAJJAR and BONNIFAIT 2005
M. E. El Najjar and P. Bonnifait (2005). "A road-matching method for precise vehicle localization using belief theory and Kalman filtering." In: *Autonomous Robots* 19.2, pp. 173–191

ESKANDARIAN 2012
A. Eskandarian (2012). *Handbook of intelligent vehicles*. Springer

FOUQUE et al. 2008
C. Fouque, P. Bonnifait, and D. Bétaille (2008). "Enhancement of global vehicle localization using navigable road maps and dead-reckoning." In: *Position, Location and Navigation Symposium, 2008 IEEE/ION*, 1286–1291

GEBRU et al. 2016
I. Gebru, X. Alameda-Pineda, F. Forbes, and R. Horaud (2016). "EM algorithms for weighted-data clustering with application to audio-visual scene analysis." In: *IEEE Trans. Pattern Anal. Machine Intell.* 38.12 (Dec. 2016), pp. 2402–2415. ISSN: 0162-8828

GENOVESE 2001
A. F. Genovese (2001). "The interacting multiple model algorithm for accurate state estimation of maneuvering targets." In: *Johns Hopkins APL technical digest* 22.4, pp. 614–623

GEORGE et al. 2012
P. George, I. Thouvenin, V. Frémont, and V. Cherfaoui (2012). "DAARIA: Driver assistance by augmented reality for intelligent automobile." In: *IEEE Intelligent Vehicles Symp. (IV)*. Alcalá de Henares, Spain, June 2012, 1043–1048

GEORGY et al. 2012
J. Georgy, A. Noureldin, and C. Goodall (2012). “Vehicle navigator using a mixture particle filter for inertial sensors/odometer/map data/GPS integration.” In: *IEEE Transactions on Consumer Electronics* 58.2, pp. 544–552

GORDON et al. 1993
N. J. Gordon, D. J. Salmond, and A. F. M. Smith (1993). “Novel approach to nonlinear/non-Gaussian Bayesian state estimation.” In: *IEE Proceedings F – Radar and Signal Processing* 140.2 (Apr. 1993), pp. 107–113. ISSN: 0956-375X

HEIN 2000
G. W. Hein (2000). “From GPS and GLONASS via EGNOS to Galileo – Positioning and Navigation in the Third Millennium.” In: *GPS Solutions* 3.4 (Apr. 2000), pp. 39–47. ISSN: 1080-5370

HERE 2015
HERE (2015). *GDF 3.0 Reference Manual v53.0*. Jan. 1, 2015

HÉRY et al. 2017
E. Héry, P. Xu, and P. Bonnifait (2017). “Along-track Localization for Cooperative Autonomous Vehicles.” In: *IEEE Intelligent Vehicles Symp. (IV)*. Redondo Beach, USA, June 2017, pp. 511–516

HOL et al. 2006
J. D. Hol, T. B. Schon, and F. Gustafsson (2006). “On resampling algorithms for particle filters.” In: *Nonlinear Statistical Signal Processing Workshop, 2006 IEEE*, pp. 79–82

HONDA MOTOR CO. 2017
Honda Motor Co. (2017). *Honda Electro Gyrocator Recognized as IEEE Milestone*. Mar. 2, 2017. URL: `http://world.honda.com/news/2017/c170302eng.html` (visited on 09/01/2017)

HU and UCHIMURA 2004
Z. Hu and K. Uchimura (2004). “Real-time data fusion on stabilizing camera pose estimation output for vision-based road navigation.” In: *Electronic Imaging 2004*, pp. 480–490

HWANG and SUNG 2013
K. Hwang and W. Sung (2013). "Load balanced resampling for real-time particle filtering on graphics processing units." In: *IEEE Transactions on Signal Processing* 61.2 (Jan. 2013), pp. 411–419

IKEA 2013
IKEA (2013). *Place IKEA furniture in your home with augmented reality*. July 26, 2013. URL: `https://www.youtube.com/watch?v=vDNzTasuYEw` (visited on 09/01/2017)

ISARD and BLAKE 1998
M. Isard and A. Blake (1998). "Condensation—conditional density propagation for visual tracking." In: *International journal of computer vision* 29.1, pp. 5–28

JOSWIG 2016
B. Joswig (2016). *Konzeption eines heuristischen Verfahrens auf Basis von Fahrzeugsensordaten zur spurgenauen Navigation*. Praktikumsbericht. Sindelfingen: DHBW, Sept. 2016

KALMAN 1960
R. E. Kalman (1960). "A new approach to linear filtering and prediction problems." In: *J. Basic Eng.* 82.1, pp. 35–45

KASTL 2017
T. Kastl (2017). "Fahrstreifengenaue Lokalisierung bei Ausfall der Umfeldsensorik." MSc thesis. Karlsruhe: Karlsruher Institut für Technologie, Sept. 2017

KRAKIWSKY et al. 1988
E. J. Krakiwsky, C. B. Harris, and R. V. Wong (1988). "A Kalman filter for integrating dead reckoning, map matching and GPS positioning." In: *IEEE Position, Location and Navigation Symp. (PLANS)*, pp. 39–46

KUHNT et al. 2016
F. Kuhnt, S. Orf, S. Klemm, and J. M. Zöllner (2016). "Lane-precise Localization of Intelligent Vehicles Using the Surrounding Object Constellation." In: *19th Int. IEEE Conf. on Intelligent Transportation Systems (ITSC)*. Rio de Janeiro, Brazil, Nov. 2016, pp. 526–533

LATEGAHN et al. 2013
H. Lategahn, J. Beck, B. Kitt, and C. Stiller (2013). "How to learn an illumination robust image feature for place recognition." In: *IEEE Intelligent Vehicles Symp. (IV)*. Gold Coast City, Australia, June 2013, pp. 285–291

LATEGAHN and STILLER 2012
H. Lategahn and C. Stiller (2012). "City GPS using stereo vision." In: *IEEE Int. Conf. on Vehicular Electronics and Safety (ICVES)*. Istanbul, Turkey, pp. 1–6

LEE et al. 2015
J.-W. Lee, C.-R. Yoon, J. Kang, B.-J. Park, and K.-H. Kim (2015). "Development of lane-level guidance service in vehicle augmented reality system." In: *17th Int. Conf. on Advanced Communication Technology (ICACT)*. Pyeongchang, Republic of Korea, July 2015, pp. 263–266

LEVINSON et al. 2007
J. Levinson, M. Montemerlo, and S. Thrun (2007). "Map-Based Precision Vehicle Localization in Urban Environments." In: *Robotics: Science and Systems*

LEVINSON and THRUN 2010
J. Levinson and S. Thrun (2010). "Robust vehicle localization in urban environments using probabilistic maps." In: *IEEE Int. Conf. Robotics and Automation (ICRA)*, pp. 4372–4378

F. LI et al. 2017
F. Li, P. Bonnifait, J. Ibanez-Guzman, and C. Zinoune (2017). "Lane-Level Map-Matching with Integrity on High-Definition Maps." In: *IEEE Intelligent Vehicles Symp. (IV)*. Redondo Beach, USA, June 2017, pp. 1176–1181

Y. LI et al. 2006
Y. Li, J. Wang, C. Rizos, P. Mumford, and W. Ding (2006). "Low-cost tightly coupled GPS/INS integration based on a nonlinear Kalman filtering design." In: *Proceedings of ION National Technical Meeting*, pp. 18–20

LIN et al. 2011
J.-H. Lin, C.-M. Lin, C.-R. Dow, and C.-Q. Wang (2011). "Design and Implement Augmented Reality for Supporting Driving Visual Guidance." In: *2nd Int. Conf. Innovations in Bio-inspired Computing and Applications*. Dec. 2011, pp. 316–319

LU et al. 2014
W. Lu, E. Seignez, S. A. Rodriguez Florez, and R. Reynaud (2014). "Lane marking based vehicle localization using particle filter and multi-kernel estimation." In: *13th Int. Conf. on Control Automation Robotics & Vision (ICARCV)*. Singapore, Dec. 2014, pp. 601–606

MATHESON 2012
R. Matheson (2012). *Buick Promotes 2013 Encore in China with Augmented Reality 'Hide & Seek' Game*. Sept. 22, 2012. URL: `http://randymatheson.com/?p=2174` (visited on 09/01/2017)

MATTERN et al. 2010
N. Mattern, R. Schubert, and G. Wanielik (2010). "High-accurate vehicle localization using digital maps and coherency images." In: *IEEE Intelligent Vehicles Symp. (IV)*. San Diego, USA, June 2010, pp. 462–469

MATTHAEI et al. 2014
R. Matthaei, G. Bagschik, and M. Maurer (2014). "Map-Relative Localization in Lane-Level Maps for ADAS and Autonomous Driving." In: *IEEE Intelligent Vehicles Symp. (IV)*. Dearborn, Michigan, USA, June 9, 2014

MAYER et al. 2014
M. Mayer, M. Aleksic, A. Bracht, B. Raichle, and S. Quell (2014). "Verfahren zum Betreiben einer Fahrerassistenzeinrichtung (German), [Method for operating driver assistance system for motor vehicle, involves detecting current driving operation related parameters, and weighing current driving operation related parameters in probability distribution calculation]." DE102013016596 A1. Apr. 17, 2014

MEINKE 2015
M. Meinke (2015). "An Optimization-Based Approach to Lane-level Map-Matching on Accurate Digital Maps." MSc thesis. Tübingen: Eberhard Karls Universität Tübingen, Sept. 2015

DAIMLER 2011
Mercedes-Benz (2011). *Mercedes-Benz Accessories: Augmented Reality Apps*. Nov. 25, 2011. URL: `https://www.youtube.com/watch?v=1P2uAOyUkgI` (visited on 08/30/2017)

MILSTEIN et al. 2002
A. Milstein, J. N. Sánchez, and E. T. Williamson (2002). "Robust global localization using clustered particle filtering." In: *Proceedings of the Eighteenth National Conference on Artificial Intelligence*. Eighteenth National Conference on Artificial Intelligence (AAAI-02). Edmonton, Alberta, Canada, Aug. 28, 2002, pp. 581–586

MITCHELL 2007
H. Mitchell (2007). "Robust Statistics." In: *Multi-Sensor Data Fusion: An Introduction*. Berlin, Heidelberg: Springer Berlin Heidelberg, pp. 155–171. ISBN: 978-3-540-71559-7

MULLONI 2012
A. Mulloni (2012). "Enhancing Handheld Navigation Systems with Augmented Reality." Institute for Computer Graphics and Vision. PhD thesis. Technische Universität Graz Graz University of Technology

NEDEVSCHI et al. 2013
S. Nedevschi, V. Popescu, R. Danescu, T. Marita, and F. Oniga (2013). "Accurate Ego-Vehicle Global Localization at Intersections Through Alignment of Visual Data With Digital Map." In: *IEEE Trans. Intell. Transp. Syst.* 14.2 (June 2013), pp. 673–687. ISSN: 1524-9050, 1558-0016

NIANTIC 2016
Niantic (2016). *Break out the sneakers and Poké Balls!* July 6, 2016. URL: `https://www.nianticlabs.com/blog/launch/` (visited on 09/01/2017)

NODA et al. 2011
M. Noda, T. Takahashi, D. Deguchi, I. Ide, H. Murase, Y. Kojima, and T. Naito (2011). "Vehicle ego-localization by matching in-vehicle camera images to an aerial image." In: *Computer Vision–ACCV 2010 Workshops*, pp. 163–173

OSKIPER et al. 2012
T. Oskiper, S. Samarasekera, and R. Kumar (2012). "Multi-sensor navigation algorithm using monocular camera, IMU and GPS for large scale augmented reality." In: *IEEE Int. Symp. Mixed and Augmented Reality (ISMAR)*, 71–80

PARK et al. 2012
J. Park, D. Lee, and J. Park (2012). "Digital map based pose improvement for outdoor Augmented Reality." In: *IEEE Int. Symp. Mixed and Augmented Reality (ISMAR)*, pp. 309–310

PETRICH et al. 2013
D. Petrich, T. Dang, D. Kasper, G. Breuel, and C. Stiller (2013). "Map-based Long Term Motion Prediction for Vehicles in Traffic Environments." In: *16th Int. IEEE Conf. on Intelligent Transportation Systems (ITSC)*. The Hague, Netherlands, Oct. 2013, pp. 2166–2172

PINK 2008
O. Pink (2008). "Visual map matching and localization using a global feature map." In: *Computer Vision and Pattern Recognition Workshops*, pp. 1–7

POPESCU et al. 2012
V. Popescu, R. Danescu, and S. Nedevschi (2012). "On-road position estimation by probabilistic integration of visual cues." In: *IEEE Intelligent Vehicles Symp. (IV)*. Alcalá de Henares, Spain, June 2012, pp. 583–589

QIN et al. 2012
B. Qin, Z. Chong, T. Bandyopadhyay, M. Ang Jr, E. Frazzoli, and D. Rus (2012). "Curb-intersection feature based monte carlo localization on urban roads." In: *IEEE Int. Conf. Robotics and Automation (ICRA)*, pp. 2640–2646

QUDDUS and VELAGA 2012
M. A. Quddus and N. R. Velaga (2012). "Enhancing Vehicle Positioning Data Through Map-Matching." In: *Handbook of Intelligent Vehicles*. Ed. by A. Eskandarian. Springer London, pp. 343–364. ISBN: 978-0-85729-084-7

RABE et al. 2015
J. Rabe, A. Danner, and C. Grünler (2015). "Map-Matching for In-Vehicle Augmented Reality Applications." In: *Virtuelle und Erweiterte Realität. 12. Workshop der GI-Fachgruppe VR/AR*. Ed. by A. Hinkenjann, J. Maiero, and R. Blach. Sankt Augustin, Germany, Sept. 2015. ISBN: 978-3-8440-3868-2

RABE et al. 2017
J. Rabe, M. Hübner, M. Necker, and C. Stiller (2017). "Ego-Lane Estimation for Downtown Lane-Level Navigation." In: *IEEE Intelligent Vehicles Symp. (IV)*. Redondo Beach, CA, USA, June 2017, pp. 1158–1163. ISBN: 978-1-5090-4803-8

RABE and JOSWIG 2016
J. Rabe and B. Joswig (2016). "Lane-Precise Navigation on Incomplete Maps." In: *2nd KuVS Expert Talk on Localization*. Ed. by M. Pelka, J. Á. Bitsch, G. Goronzy, H. Hellbrück, and K. Wehrle. Vol. AIB-2016-05. ISSN 0935–3232. Lübeck, Germany, July 15, 2016, pp. 27–28

RABE et al. 2016a
J. Rabe, M. Meinke, M. Necker, and C. Stiller (2016a). "Lane-level map-matching based on optimization." In: *19th Int. IEEE Conf. on Intelligent Transportation Systems (ITSC)*. Rio de Janeiro, Brazil, Nov. 2016, pp. 1155–1160

RABE et al. 2016b
J. Rabe, M. Necker, and C. Stiller (2016b). "Ego-lane estimation for lane-level navigation in urban scenarios." In: *IEEE Intelligent Vehicles Symp. (IV)*. Gothenburg, Sweden, June 2016, pp. 896–901

RABE and STILLER 2017
J. Rabe and C. Stiller (2017). "Robust Particle Filter for Lane-Precise Localization." In: *IEEE Int. Conf. on Vehicular Electronics and Safety (ICVES)*. Vienna, Austria, June 2017, pp. 127–132. ISBN: 978-1-5090-5676-7

RAJAMANI 2012
R. Rajamani (2012). *Vehicle Dynamics and Control*. Mechanical Engineering Series. Springer. ISBN: 978-1-4614-1433-9

RAO et al. 2014a
Q. Rao, C. Grünler, M. Hammori, and S. Chakraborty (2014a). "Design methods for augmented reality in-vehicle infotainment systems." In: *Proceedings of the 51st Annual Design Automation Conference*. 2014, pp. 1–6

RAO et al. 2014b
Q. Rao, T. Tropper, C. Grünler, M. Hammori, and S. Chakraborty (2014b). "AR-IVI—Implementation of In-Vehicle Augmented Reality." In: *IEEE Int. Symp. Mixed and Augmented Reality (ISMAR)*. Munich, Germany, Sept. 2014, pp. 3–8

SALARIAN and ANSARI 2015
M. Salarian and R. Ansari (2015). "Efficient refinement of GPS-based localization in urban areas using visual information and sensor parameter." In: *CoRR* arXiv:1502.00319

SALAS-MORENO et al. 2014
R. F. Salas-Moreno, B. Glocken, P. H. Kelly, and A. J. Davison (2014). "Dense planar SLAM." In: *17th IEEE Int. Symp. Mixed and Augmented Reality (ISMAR)*, pp. 157–164

SCHALL et al. 2009
G. Schall, D. Wagner, G. Reitmayr, E. Taichmann, M. Wieser, D. Schmalstieg, and B. Hofmann-Wellenhof (2009). "Global pose estimation using multi-sensor fusion for outdoor augmented reality." In: *8th IEEE Int. Symp. Mixed and Augmented Reality (ISMAR)*, pp. 153–162

SCHÖPS et al. 2014
T. Schöps, J. Engel, and D. Cremers (2014). "Semi-dense visual odometry for AR on a smartphone." In: *Mixed and Augmented Reality (ISMAR), 2014 IEEE International Symposium on*, pp. 145–150

SCHRAMM et al. 2014
D. Schramm, M. Hiller, and R. Bardini (2014). *Vehicle Dynamics: Modeling and Simulation*. Springer Berlin Heidelberg. ISBN: 978-3-540-36045-2

SCHREIBER et al. 2013
M. Schreiber, C. Knöppel, and U. Franke (2013). "LaneLoc: Lane marking based localization using highly accurate maps." In: *IEEE Intelligent Vehicles Symp. (IV)*. Gold Coast City, Australia, June 2013, pp. 449–454

SCHÜLE 2015
F. Schüle (2015). "Kartenbasierte Straßenverlaufsprädiktion zur Sichtunterstützung bei Nacht." PhD thesis. Ulm: Universität Ulm

SCHÜLE et al. 2013
F. Schüle, R. Schweiger, and K. Dietmayer (2013). "Augmenting night vision video images with longer distance road course information." In: *IEEE Intelligent Vehicles Symp. (IV)*. Gold Coast City, Australia, June 2013, 1233–1238

SCHÜLE et al. 2011
F. Schüle, R. Schweiger, O. Löhlein, and K. Dietmayer (2011). "Vehicle Positioning on a Digital Map for Road Course Prediction." In: *First International Symposium on Future Active Safety Technology toward zero-traffic-accident*. FAST-zero. Tokyo, Japan, Sept. 5, 2011

SELLOUM et al. 2009
A. Selloum, D. Betaille, E. Le Carpentier, and F. Peyret (2009). "Lane level positioning using particle filtering." In: *12th Int. IEEE Conf. on Intelligent Transportation Systems (ITSC)*. Beijing, China, Oct. 2009, pp. 1–6

SHEN 2013
K. Shen (2013). "Augmented Navigation. Verschmelzung von Routen, Karte und Realität." In: *Automobiltechnische Zeitschrift* 115.5

SIEMENS AG 2005
Siemens AG (2005). *Augmented Reality Navigation in the shuttle bus*. Sept. 30, 2005

B. SIMON 2017

B. Simon (2017). *Straight to your destination: The "EVA" electronic navigation guide for drivers.* June 19, 2017. URL: http://blog.bosch.com/history/en/2017/06/19/straight-to-your-destination-the-eva-electronic-navigation-guide-for-drivers/ (visited on 09/01/2017)

C. R. SIMON 2012

C. R. Simon (2012). "Straßenumfeldmodellierung durch Fusion und Interpretation von Sensordaten zur spurgenauen Positionierung des Fahrzeugs." PhD thesis. München: Technische Universität München

SVENSSON and SÖRSTEDT 2016

D. Svensson and J. Sörstedt (2016). "Ego lane estimation using vehicle observations and map information." In: *IEEE Intelligent Vehicles Symp. (IV)*. Gothenburg, Sweden, June 2016, pp. 909–914

SZOTTKA 2013

I. Szottka (2013). "Particle filtering for lane-level map-matching at road bifurcations." In: *16th Int. IEEE Conf. on Intelligent Transportation Systems (ITSC)*. The Hague, Netherlands, Oct. 2013, pp. 154–159

TAO et al. 2013

Z. Tao, P. Bonnifait, V. Fremont, and J. Ibanez-Guzman (2013). "Lane marking aided vehicle localization." In: *16th Int. IEEE Conf. on Intelligent Transportation Systems (ITSC)*, pp. 1509–1515

TESSIER et al. 2007

C. Tessier, C. Debain, R. Chapuis, and F. Chausse (2007). "Simultaneous landmarks detection and data association in noisy environment for map aided localization." In: *IEEE/RSJ Int. Conf. Intelligent Robots and Systems (IROS)*. San Diego, USA, pp. 1396–1401

THRUN et al. 2005

S. Thrun, W. Burgard, and D. Fox (2005). *Probabilistic Robotics*. Intelligent robotics and autonomous agents. MIT Press. ISBN: 978-0-262-20162-9

TOLEDO-MOREO et al. 2010
R. Toledo-Moreo, D. Betaille, and F. Peyret (2010). "Lane-Level Integrity Provision for Navigation and Map Matching With GNSS, Dead Reckoning, and Enhanced Maps." In: *IEEE Trans. Intell. Transp. Syst.* 11.1 (Mar. 2010), pp. 100–112. ISSN: 1524-9050, 1558-0016

TOLEDO-MOREO et al. 2009
R. Toledo-Moreo, D. Betaille, F. Peyret, and J. Laneurit (2009). "Fusing GNSS, Dead-Reckoning, and Enhanced Maps for Road Vehicle Lane-Level Navigation." In: *IEEE Journal of Selected Topics in Signal Processing* 3.5 (Oct. 2009), pp. 798–809. ISSN: 1932-4553

TOMTOM 2013
TomTom Global Content BV and TomTom North America (2013). *MultiNet ® Standard Data Specifications 3.6.3*. Sept. 12, 2013

TOMTOM 2016
TomTom International BV (2016). *Highly Detailed Map*. URL: http://automotive.tomtom.com/en/highly-automated-driving/highly-detailed-map (visited on 08/17/2016)

TÖNNIS 2008
M. Tönnis (2008). "Towards Automotive Augmented Reality." Fakultät für Informatik Prof. Gudrun Klinker. PhD thesis. Technische Universität München

VACEK et al. 2006
S. Vacek, S. Bergmann, U. Mohr, and R. Dillmann (2006). "Fusing image features and navigation system data for augmenting guiding information displays." In: *IEEE Int. Conf. Multisensor Fusion and Integration for Intelligent Systems (MFI)*, pp. 323–328

VAN DER MERWE 2004
R. Van Der Merwe (2004). "Sigma-point Kalman filters for probabilistic inference in dynamic state-space models." PhD thesis. Oregon Health & Science University

VAN HAMME et al. 2013
D. Van Hamme, P. Veelaert, and W. Philips (2013). "Lane identification based on robust visual odometry." In: *16th Int. IEEE Conf. on Intelligent Transportation Systems (ITSC)*. The Hague, Netherlands, Oct. 2013, pp. 1179–1183

WALLENTOWITZ and REIF 2010
H. Wallentowitz and K. Reif (2010). *Handbuch Kraftfahrzeugelektronik: Grundlagen - Komponenten - Systeme - Anwendungen*. ATZ/MTZ-Fachbuch. Vieweg+Teubner Verlag. ISBN: 978-3-8348-0700-7

WANG et al. 2011
C. Wang, Z. Hu, and K. Uchimura (2011). "A novel lane detection approach fusion by vehicle localization." In: *World Congr. Intelligent Control and Automation (WCICA)*, pp. 1218–1223

WILCOX 2001
R. R. Wilcox (2001). *Fundamentals of Modern Statistical Methods: Substantially Improving Power and Accuracy*. New York, NY: Springer New York. ISBN: 978-1-4757-3522-2

XU 2007
G. Xu (2007). *GPS: Theory, Algorithms and Applications*. Springer. ISBN: 978-3-540-72715-6. URL: https://books.google.de/books?id=peYFZ69HqEsC

YONG and CHENG 2004
J.-H. Yong and F. Cheng (2004). "Geometric Hermite curves with minimum strain energy." In: *Computer Aided Geometric Design* 21.3, pp. 281–301

ZIEGLER et al. 2014
J. Ziegler, P. Bender, M. Schreiber, H. Lategahn, T. Strauss, C. Stiller, T. Dang, U. Franke, N. Appenrodt, C. G. Keller, E. Kaus, R. G. Herrtwich, C. Rabe, D. Pfeiffer, F. Lindner, F. Stein, F. Erbs, M. Enzweiler, C. Knöppel, J. Hipp, M. Haueis, M. Trepte, C. Brenk, A. Tamke, M. Ghanaat, M. Braun, A. Joos, H. Fritz, H. Mock, M. Hein, and E. Zeeb (2014). "Making Bertha Drive–An Autonomous Journey on a Historic Route." In: *IEEE Intell. Transp. Syst. Mag.* 6.2, pp. 8–20. ISSN: 1939-1390

Publications by the Author

J. Rabe, A. Danner, and C. Grünler (2015a). "Map-Matching for In-Vehicle Augmented Reality Applications." In: *Virtuelle und Erweiterte Realität. 12. Workshop der GI-Fachgruppe VR/AR*. Ed. by A. Hinkenjann, J. Maiero, and R. Blach. Sankt Augustin, Germany, Sept. 2015. ISBN: 978-3-8440-3868-2

J. Rabe, M. Hübner, M. Necker, and C. Stiller (2017). "Ego-Lane Estimation for Downtown Lane-Level Navigation." In: *IEEE Intelligent Vehicles Symp. (IV)*. Redondo Beach, CA, USA, June 2017, pp. 1158–1163. ISBN: 978-1-5090-4803-8

J. Rabe and B. Joswig (2016). "Lane-Precise Navigation on Incomplete Maps." In: *2nd KuVS Expert Talk on Localization*. Ed. by M. Pelka, J. Á. Bitsch, G. Goronzy, H. Hellbrück, and K. Wehrle. Vol. AIB-2016-05. ISSN 0935–3232. Lübeck, Germany, July 15, 2016, pp. 27–28

J. Rabe, M. Meinke, M. Necker, and C. Stiller (2016a). "Lane-level map-matching based on optimization." In: *19th Int. IEEE Conf. on Intelligent Transportation Systems (ITSC)*. Rio de Janeiro, Brazil, Nov. 2016, pp. 1155–1160

J. Rabe, M. Necker, and C. Stiller (2016b). "Ego-lane estimation for lane-level navigation in urban scenarios." In: *IEEE Intelligent Vehicles Symp. (IV)*. Gothenburg, Sweden, June 2016, pp. 896–901

J. Rabe, S. Quell, M. Necker, and C. Stiller (2015b). "Lane-Level Localization on Lanelet Maps Using Production Vehicle Sensors." In: *1st KuVS Expert Talk on Localization*. Ed. by M. Pelka, J. Á. Bitsch, H. Hellbrück, and K. Wehrle. Vol. AIB-2015-08. ISSN 0935–3232. Lübeck, Germany, Apr. 24, 2015, 25–27

J. Rabe and C. Stiller (2017). "Robust Particle Filter for Lane-Precise Localization." In: *IEEE Int. Conf. on Vehicular Electronics and Safety (ICVES)*. Vienna, Austria, June 2017, pp. 127–132. ISBN: 978-1-5090-5676-7

Patent Applications

A. Danner, R. Janke, J. Rabe, and C. Grünler (2016). "Verfahren zum Bestimmen einer aktuellen Position eines Kraftfahrzeugs auf einer Fahrbahn sowie Positionsbestimmungsvorrichtung (German), [A method for determining a current position of a vehicle on a roadway and position determining device]." DE102016001269 A1. Nov. 8, 2016

M. Necker and J. Rabe (2015). "Verfahren zur Positionsbestimmung eines Fahrzeugs (German) [A method for determining the position of a vehicle]." DE102015001360 A1. Aug. 20, 2015

J. Rabe (2017a). "Verfahren zur Lokalisierung eines Fahrzeugs bezüglich einer Fahrbahn (German) [A method for locating a vehicle relative to a road surface]." DE102017001307 A1. Oct. 19, 2017

— (2017b). "Verfahren zur Positionsbestimmung eines Fahrzeugs (German) [A method for determining the position of a vehicle]." DE102016011366 A1. June 4, 2017

J. Rabe and B. Joswig (2017). "Verfahren zum Unterstützen eines Fahrers eines Kraftfahrzeugs beim Manövrieren des Kraftfahrzeugs sowie Fahrerassistenzsystem (German) [Method for assisting a driver of a motor vehicle in maneuvering of the motor vehicle and driver assistance system]." DE102016008382 A1. Feb. 2, 2017

J. Rabe, M. Necker, and S. Quell (2015). "Verfahren zum Betreiben einer Fahrerassistenzeinrichtung eines Kraftwagens, sowie Fahrerassistenzeinrichtung für einen Kraftwagen (German), [A method for operating a driver assistance device of a motor vehicle, and driver assistance device for a motor vehicle]." DE102015004676 A1. Mar. 12, 2015

Supervised Theses

A. Danner (2015). *Anforderungsspezifische Map-Matching Erweiterung für ein Augmented Reality System im Fahrzeug*. Praktikumsbericht. Sindelfingen: DHBW, June 2015

R. Janke (2015). *Map-Matching und Koppelortung für ein Augmented Reality-System im Fahrzeug*. Praktikumsbericht. Sindelfingen: DHBW, Sept. 2015

B. Joswig (2016). *Konzeption eines heuristischen Verfahrens auf Basis von Fahrzeugsensordaten zur spurgenauen Navigation*. Praktikumsbericht. Sindelfingen: DHBW, Sept. 2016

T. Kastl (2017). "Fahrstreifengenaue Lokalisierung bei Ausfall der Umfeldsensorik." MSc thesis. Karlsruhe: Karlsruher Institut für Technologie, Sept. 2017

M. Meinke (2015). "An Optimization-Based Approach to Lane-level Map-Matching on Accurate Digital Maps." MSc thesis. Tübingen: Eberhard Karls Universität Tübingen, Sept. 2015

M. Nichting (2014). *Eigenfahrstreifendetektion für Navigation mit Augmented Reality*. Praktikumsbericht. Sindelfingen: DHBW, Sept. 2014

E. Schütz (2015). *Erstellung eines spurgenauen Augmented Reality Navigationsteppichs*. Praktikumsbericht. Sindelfingen: DHBW, June 2015

E. Ülger (2015). "Einfluss der Qualität von Kartendaten auf die Genauigkeit der Positionsbestimmung bei Augmented Reality." MSc thesis. Tübingen: Eberhard Karls Universität Tübingen, Sept. 2015

A. Uzun (2015). *Realisierung eines verbesserten spurgenauen Augmented Reality Navigationsteppichs*. Praktikumsbericht. Sindelfingen: DHBW, Sept. 2015

Schriftenreihe
Institut für Mess- und Regelungstechnik
Karlsruher Institut für Technologie
(1613-4214)

Band 001	Hans, Annegret **Entwicklung eines Inline-Viskosimeters auf Basis eines magnetisch-induktiven Durchflussmessers.** 2004 ISBN 3-937300-02-3
Band 002	Heizmann, Michael **Auswertung von forensischen Riefenspuren mittels automatischer Sichtprüfung.** 2004 ISBN 3-937300-05-8
Band 003	Herbst, Jürgen **Zerstörungsfreie Prüfung von Abwasserkanälen mit Klopfschall.** 2004 ISBN 3-937300-23-6
Band 004	Kammel, Sören **Deflektometrische Untersuchung spiegelnd reflektierender Freiformflächen.** 2005 ISBN 3-937300-28-7
Band 005	Geistler, Alexander **Bordautonome Ortung von Schienenfahrzeugen mit Wirbelstrom-Sensoren.** 2007 ISBN 978-3-86644-123-1
Band 006	Horn, Jan **Zweidimensionale Geschwindigkeitsmessung texturierter Oberflächen mit flächenhaften bildgebenden Sensoren.** 2007 ISBN 978-3-86644-076-0

Die Bände sind unter www.ksp.kit.edu als PDF frei verfügbar oder als Druckausgabe bestellbar.

Band 007	Hoffmann, Christian **Fahrzeugdetektion durch Fusion monoskopischer Videomerkmale.** 2007 ISBN 978-3-86644-139-2
Band 008	Dang, Thao **Kontinuierliche Selbstkalibrierung von Stereokameras.** 2007 ISBN 978-3-86644-164-4
Band 009	Kapp, Andreas **Ein Beitrag zur Verbesserung und Erweiterung der Lidar-Signalverarbeitung für Fahrzeuge.** 2007 ISBN 978-3-86644-174-3
Band 010	Horbach, Jan **Verfahren zur optischen 3D-Vermessung spiegelnder Oberflächen.** 2008 ISBN 978-3-86644-202-3
Band 011	Böhringer, Frank **Gleisselektive Ortung von Schienenfahrzeugen mit bordautonomer Sensorik.** 2008 ISBN 978-3-86644-196-5
Band 012	Xin, Binjian **Auswertung und Charakterisierung dreidimensionaler Messdaten technischer Oberflächen mit Riefentexturen.** 2009 ISBN 978-3-86644-326-6
Band 013	Cech, Markus **Fahrspurschätzung aus monokularen Bildfolgen für innerstädtische Fahrerassistenzanwendungen.** 2009 ISBN 978-3-86644-351-8
Band 014	Speck, Christoph **Automatisierte Auswertung forensischer Spuren auf Patronenhülsen.** 2009 ISBN 978-3-86644-365-5

Band 015	Bachmann, Alexander **Dichte Objektsegmentierung in Stereobildfolgen.** 2010 ISBN 978-3-86644-541-3
Band 016	Duchow, Christian **Videobasierte Wahrnehmung markierter Kreuzungen mit lokalem Markierungstest und Bayes'scher Modellierung.** 2011 ISBN 978-3-86644-630-4
Band 017	Pink, Oliver **Bildbasierte Selbstlokalisierung von Straßenfahrzeugen.** 2011 ISBN 978-3-86644-708-0
Band 018	Hensel, Stefan **Wirbelstromsensorbasierte Lokalisierung von Schienenfahrzeugen in topologischen Karten.** 2011 ISBN 978-3-86644-749-3
Band 019	Carsten Hasberg **Simultane Lokalisierung und Kartierung spurgeführter Systeme.** 2012 ISBN 978-3-86644-831-5
Band 020	Pitzer, Benjamin **Automatic Reconstruction of Textured 3D Models.** 2012 ISBN 978-3-86644-805-6
Band 021	Roser, Martin **Modellbasierte und positionsgenaue Erkennung von Regentropfen in Bildfolgen zur Verbesserung von videobasierten Fahrerassistenzfunktionen.** 2012 ISBN 978-3-86644-926-8

Band 022 Loose, Heidi
Dreidimensionale Straßenmodelle für Fahrerassistenzsysteme auf Landstraßen. 2013
ISBN 978-3-86644-942-8

Band 023 Rapp, Holger
Reconstruction of Specular Reflective Surfaces using Auto-Calibrating Deflectometry. 2013
ISBN 978-3-86644-966-4

Band 024 Moosmann, Frank
Interlacing Self-Localization, Moving Object Tracking and Mapping for 3D Range Sensors. 2013
ISBN 978-3-86644-977-0

Band 025 Geiger, Andreas
Probabilistic Models for 3D Urban Scene Understanding from Movable Platforms. 2013
ISBN 978-3-7315-0081-0

Band 026 Hörter, Marko
Entwicklung und vergleichende Bewertung einer bildbasierten Markierungslichtsteuerung für Kraftfahrzeuge. 2013
ISBN 978-3-7315-0091-9

Band 027 Kitt, Bernd
Effiziente Schätzung dichter Bewegungsvektorfelder unter Berücksichtigung der Epipolargeometrie zwischen unterschiedlichen Ansichten einer Szene. 2013
ISBN 978-3-7315-0105-3

Band 028 Lategahn, Henning
Mapping and Localization in Urban Environments Using Cameras. 2013
ISBN 978-3-7315-0135-0

Band 029	Tischler, Karin **Informationsfusion für die kooperative Umfeldwahrnehmung vernetzter Fahrzeuge.** 2014 ISBN 978-3-7315-0166-4
Band 030	Schmidt, Christian **Fahrstrategien zur Unfallvermeidung im Straßenverkehr für Einzel- und Mehrobjektszenarien.** 2014 ISBN 978-3-7315-0198-5
Band 031	Firl, Jonas **Probabilistic Maneuver Recognition in Traffic Scenarios.** 2014 ISBN 978-3-7315-0287-6
Band 032	Schönbein, Miriam **Omnidirectional Stereo Vision for Autonomous Vehicles.** 2015 ISBN 978-3-7315-0357-6
Band 033	Nicht erschienen
Band 034	Liebner, Martin **Fahrerabsichtserkennung und Risikobewertung für warnende Fahrerassistenzsysteme.** 2016 ISBN 978-3-7315-0508-2
Band 035	Ziegler, Julius **Optimale Trajektorienplanung für Automobile.** 2017 ISBN 978-3-7315-0553-2
Band 036	Harms, Hannes **Genauigkeitsuntersuchung von binokularen Normalenvektoren für die Umfeldwahrnehmung.** 2017 ISBN 978-3-7315-0628-7

Band 037 Ruhhammer, Christian
Inferenz von Kreuzungsinformationen aus Flottendaten. 2017
ISBN 978-3-7315-0721-5

Band 038 Stein, Denis
Mobile laser scanning based determination of railway network topology and branching direction on turnouts. 2018
ISBN 978-3-7315-0743-7

Band 039 Yi, Boliang
Integrated Planning and Control for Collision Avoidance Systems. 2018
ISBN 978-3-7315-0785-7

Band 040 Schwarze, Tobias
Compact Environment Modelling from Unconstrained Camera Platforms. 2018
ISBN 978-3-7315-0801-4

Band 041 Knorr, Moritz
Self-Calibration of Multi-Camera Systems for Vehicle Surround Sensing. 2018
ISBN 978-3-7315-0765-9

Band 042 Rabe, Johannes
Lane-Precise Localization with Production Vehicle Sensors and Application to Augmented Reality Navigation. 2018
ISBN 978-3-7315-0854-0